AN INTRODUCTION TO

SPECIAL RELATIVITY

AND ITS APPLICATIONS

AN INTRODUCTION TO SPECIAL RELATIVITY AND ITS APPLICATIONS

F N H Robinson

The Clarendon Laboratory
Oxford

World Scientific
Singapore • New Jersey • London • Hong Kong

Published by

World Scientific Publishing Co. Pte. Ltd.
P O Box 128, Farrer Road, Singapore 912805
USA office: Suite 1B, 1060 Main Street, River Edge, NJ 07661
UK office: 57 Shelton Street, Covent Garden, London WC2H 9HE

AN INTRODUCTION TO SPECIAL RELATIVITY AND ITS APPLICATIONS

ISBN: 981-02-2499-0

This book is printed on acid-free paper.

Printed in Singapore by Uto-Print

PREFACE

Many years ago I began to write a book about special relativity and its relation to classical mechanics and electromagnetism. The first draft got longer and longer and when I persuaded my colleague Dr.M.G.Bowler to read it, he thought that it was so long that he could not imagine who might want to read it. I rather agreed with him and set the whole thing aside. Years later, as an examiner, I found that although most students could manipulate the formulae of special relativity, very few of them could explain, even qualitatively, where the notion of a variable inertial mass $m_0/\sqrt{(1-v^2/c^2)}$ came from. It then seemed to me that there might be room for an elementary book on special relativity that emphasised the basic physical ideas, their consequences and their applications other than in particle physics, where there are already two excellent texts by Hagedorn and Muirhead (see bibliography).

My plan is to give a straightforward account, using only elementary mathematics, of the basis of special relativity, and then to stress its important consequences and those aspects of relativity relevant to physics and science in general.

I share Muirhead's view that most books on relativity over-emphasise the "rods and clocks" aspects of the subject. I have a dim memory of one book that asked me to visualise the whole of space being strewn with clocks much, I suppose, as a sheet of graph paper is strewn with millimetre squares! This emphasis probably derives from the time, many years ago, when there were no reproducible atomic clocks and a great deal of circumlocution was needed to explain how identical but independent clocks and measuring rods could exist in two systems in rapid relative motion. The result is usually to create the impression that understanding relativity means abandoning all common-sense notions about space and time. I hope to show that this is false, and that the entire theory of special relativity can easily be developed from a few simple ideas even if, eventually, it has some rather bizarre consequences.

The whole aim of special relativity is to reconcile the notions implicit in Newton's first law of motion (the basis of classical dynamics) with the universal invariance of the velocity of light in vacuum (a direct consequence of Maxwell's electromagnetism). It does this by changing how time intervals and spatial distances between two events are related when they are given in two frames of reference in relative motion. This is all encapsulated in a simple formula, the Lorentz transformation, and this formula is the keystone of special relativity. The first two chapters explain how this comes about.

In deriving the Lorentz transformation I first explain why it has to be linear, something that seems to be ignored by every author except Fock (1959). Thereafter I arrange the steps in the derivation to show how much of it depends only on assuming that there are no preferred positions or directions in space, and that all instants in time are equivalent. This makes it possible to see how one further assumption, that there is a universal limiting velocity c, leads to the Lorentz rather than the classical Galilean transformation. To agree with Maxwell's equations and experiment c must, of course, be the velocity of light.

Chapter 3 describes several of the optical and kinematic effects of relativity. Some, such as time dilation, seem strange when we try to reconcile them with those naive ideas about time and space that arise from our experience of slow motion, slow, that is, compared with the velocity of light which is about a billion (10^9) km/hr. These effects are, nevertheless, confirmed by numerous experiments.

The most important consequence of special relativity is, however, not these kinematic effects but its effect on dynamics. It alters the conservation laws for momentum, mass and energy and the response of particles to forces.

Chapter 4 reviews classical dynamics in preparation for relativistic dynamics in Chapters 5 and 6. By the end of Chapter 6 the reader will have met most of the simple kinematic and dynamical effects of relativity, and thus understand the origin of the relation $\mathbf{p} = \gamma m_0 \mathbf{v}$, and the significance of the famous formula $E=mc^2$.

Up to this point only the simplest mathematics has been used but from now on slightly greater demands begin to be made on the reader's mathematical knowledge. This might therefore be a good place to end a first course in special relativity.

Further important consequences of relativity that should form part of a physicist's general knowledge are the new insights that it provides into the structure of electromagnetism, and the way it modifies the principle of least action, the branch of classical mechanics leading most directly to quantum mechanics.

At this stage it is becoming clear that we need a better, more compact notation and so 4-vectors are introduced in Chapter 7. Since students find it easier to follow, we adopt the Minkowski notation in which an event is located at x_1,x_2,x_3,x_4 with x_4=ict, rather than introduce covariant and contravariant components and a metric tensor. The Minkowski notation would present problems if we intended to pursue relativistic wave mechanics, where i, the square root of -1, has already been pre-empted for another purpose. The use of 4-vectors allows us to write equations in a form that makes their agreement with the principle of relativity immediately obvious. Since both angular momentum and electromagnetism involve the vector product, we also need 4-tensors.

The use of 4-dimensional notation has been postponed to this late stage, to emphasise that it is merely a book-keeping aid, and a convenience. It has no more to do with the physics of relativity than the use of complex numbers has to do with the physics of A.C. circuits. A good deal of loose talk about a 4-dimensional world is caused by assuming that its physics is determined by notation.

Electromagnetism is treated in Chapters 8 and 9. Since we already know that relativity is consistent with classical electromagnetism, it can lead to nothing fundamentally new, although its effect on dynamics will alter the motion of charged particles in response to fields. Relativity can, however, lead to new and more concise ways of deriving classical results, for example, the treatment of synchrotron radiation in #9.8 is very much simpler than the classical treatment using the Liénard-Wiechert potentials.

Chapter 10 is partly about how the energy-momentum tensor combines fields, particles and continuous media in a single Lorentz invariant formulation yielding several general results, such as the centre of mass theorem. This chapter also discusses how relativity affects angular momentum, and its separation into orbital and spin components.

Chapter 11, on the principle of least action and the Hamiltonian ends with a sketch of how combining special relativity with quantum mechanics leads to electron spin and thus to the periodic table of the elements, to chemistry and to biology. As our existence depends on the chemistry of the elements, we may regard this as a particularly important consequence of relativity.

A few mainly mathematical topics, and also some aspects of the linearity of the Lorentz transformation, have been relegated to an appendix. The appendix ends with a brief section, of a mainly epistemological nature, and a table of physical constants.

There are problems at the end of each chapter and answers towards the end of the book.

References to authors quoted in the text are collected in alphabetical order at the end of the book, together with a short bibliography.

TABLE of CONTENTS

CHAPTER 1

INTRODUCTION

In thinking about the laws of physics we implicitly assume, without further discussion, that these laws are the same everywhere (space is homogeneous), that there are no preferred directions in space (space is isotropic) and also that they are the same at all times. It cannot therefore matter where we take the origin of our coordinates, how we orient their axes, or when we set our clocks to zero, the laws of physics will still be expressed by the same formulae. Classical, Galilean relativity further asserts that all velocity is only relative so that there is no universal frame of reference at absolute rest. Although acceleration has an absolute significance, the laws of dynamics are the same when expressed in terms of two frames of reference which differ only by a uniform relative velocity. An object dropped in a moving train falls vertically to the floor, though relative to the ground it is, of course, moving at an angle to the vertical.

Maxwell, though he later dropped the idea, initially treated electromagnetism as an effect occurring in a tangible medium, the æther, an idea supported by Bradley's (1728) observation of stellar aberration (see # 3.4). The identification of light as an electromagnetic wave by Maxwell and Hertz led to a number of ingenious experiments designed to detect the motion of the earth through the æther as it pursued its orbit around the sun at about 10^{-4} the velocity of light c. The null results of these experiments, especially those of Michelson and Morley (1887) (see #3.5) confirmed the expectation, inherent in Maxwell's field equations, that the velocity of light would be unaffected by the earth's motion relative to any hypothetical medium or æther at absolute rest, a result that eventually led to the abandonment of the concept of the æther. This behaviour of light is very different from sound where, if s is the velocity of sound in still air, the velocity in air flowing with a velocity v from a stationary source towards a stationary observer is $s+v$.

Maxwell's equations also predict that the velocity of light should be independent of the velocity of the source. Observations of pulsating binary X-ray stars have confirmed this to an accuracy of 2 parts in 10^9 (Brecher 1977).

By 1900 many physicists and mathematicians, notably Poincaré (1901,1904) had concluded that the notion of absolute rest was as little relevant to electromagnetism as it was to dynamics, indeed, as Lorentz realised, Maxwell's equations themselves predict that if the distribution of charge at $\mathbf{r}$ alters at time t, the effect will not be felt at $\mathbf{R}$ until a time $T=t+|\mathbf{R}-\mathbf{r}|/c$ and that this result is independent of any motion of the source or the observer.

Beginning in 1899, Lorentz sought a transformation of coordinates from one frame of reference to another frame in uniform relative motion, that would leave the form of Maxwell's equations, and thus their physical consequences (including the invariance of c the velocity of light) unaltered. Published in 1903, the **Lorentz Transformation** is the central formula of special relativity; it gives mathematical form to the idea that absolute rest has no meaning in any context. As Poincaré remarked, it has three very important consequences for the rest of physics. It predicts that moving clocks run slow, that no signal velocity can exceed c and that, if we are to preserve conservation of momentum, the relation $\mathbf{p}=m\mathbf{v}$ between mass m, velocity $\mathbf{v}$ and momentum $\mathbf{p}$ must be replaced by $\mathbf{p}=m\mathbf{v}/(1-v^2/c^2)^{\frac{1}{2}}$, a change that eventually leads to the celebrated formula $E=mc^2$. The very abstract nature of Lorentz's derivation, together with its strange consequences undermining both a deeply held intuitive notion about time and a basic law of dynamics, understandably did nothing to popularise the relativity theory of Lorentz and Poincaré.

A little later, Einstein (1905) approached the consequences of the non-existence of the æther in a much simpler way. He first of all noted the primacy amongst the laws of physics of Newton's first law of motion:- "a free particle moves along a straight line with a constant velocity". Without this law we cannot define mass, momentum, force, work or energy, i.e any of the key concepts of dynamics. This then led him to consider <u>inertial</u> frames and coordinate systems in which a free particle is either at rest or moves uniformly. The only possible difference between Cartesian coordinate systems in two distinct inertial frames, apart from a rotation of the axes or a displacement of the origins is that they be in uniform relative motion. Finally, instead of discussing all of electromagnetism as Lorentz had done, Einstein simply insisted that the velocity of light *c* should be the same in all inertial frames of reference.

The conceptual simplicity and mathematical elegance of Einstein's treatment, even though it could only lead to the same bizarre conclusions about time and dynamics, made it more acceptable to most physicists. His attempts to give an operational definition of equivalent time and length scales in different frames of reference were however less happy, largely because, when he was writing, the standard of time was based on astronomic measurements and the standard of length on a block of metal kept near Paris. Nowadays, when the standard of time is based on copies of Essen's reproducible atomic clock (Essen & Parry 1958) and the standard of length on the standard of time and a defined value for *c*, it is much easier to see what is meant by equivalent units of length or time at different places and times, and in frames in relative motion. Although we have been talking about clocks and standards of length in different inertial frames, or at different places in the same frame, this does not mean that we envisage the actual existence of atomic clocks and interferometers everywhere in space, any more than we expect to find a pair of surveyor's chains at every point on the ground that corresponds to a grid intersection on a map. Because we are shortly going to discover some odd things about the time and place attributed to the same event in two different inertial frames in relative motion, we are just being pedantic about what we mean by a statement such as 'I set my watch by the chimes of Big Ben on TV'.

As well as running at the same rate clocks need to be set, and the synchronisation of clocks at different places A and B within a single inertial frame of reference is to be regarded as the result of a procedure which relies on the universal constancy of the velocity of light. If a pulse of light is sent from A to B when the clock at A registers t_1, reaching B as the clock at B registers t_2 and is immediately returned to A, reaching A precisely as the clock at A registers t_3, then if $t_2 = \frac{1}{2}(t_1 + t_3)$ the clocks are synchronised. If this all sounds rather contrived try the problem at the end of this chapter.

Event in relativity has a rather specialised meaning and refers to something (such as the disintegration of a muon, a collision between two protons or the emission of a flash of light) that occurs at a *definite place and time*. The Lorentz transformation relates the different coordinates and time ascribed to a *single event* in two different coordinate systems which, even if they are based in two different inertial frames, use the same standards of length and time.

The remarkable thing about relativity is that these few obvious ideas about the isotropy of space and the homogeneity of space and time, the notion that absolute rest has no meaning and that the velocity of light is the same in all inertial frames should lead to a theory with profound consequences throughout science. The combined effect of these few simple notions is to change the fundamental laws of dynamics in a way that has repercussions throughout physics, engineering, chemistry and biology.

It is interesting that Lorentz's great paper '*Electromagnetic Phenomena in a System Moving with any Velocity Less than that of Light*' and Einstein's '*On the Electrodynamics of Moving Bodies*', should have such similar titles, and yet have reached their final conclusions in such different ways.

Although kinematic effects such as time dilation may seem the most novel, it is the dynamic effects that are the most important; they determine the structure of the elementary particles and, by requiring electrons to have an intrinsic spin $h/4\pi$ (Dirac 1930), they determine the nature of the chemical elements and thus the whole of biology.

Problem

1.1 Two power stations A and B operating at 50 Hz are 300 km apart and are joined by two independent cables. What will happen if A attempts to synchronise with B using one of the cables and B attempts to synchronise with A using the other?

CHAPTER 2

THE LORENTZ TRANSFORMATION

2.1 Vectors and Coordinates

In vector analysis we express relations between directed quantities e.g. the relation $\mathbf{F}=m\mathbf{a}$ between force mass and acceleration or the relation $dW=\mathbf{F}.d\mathbf{r}$ between work done, force and displacement in a way that is independent of any coordinate system. Thus if the vertices of a triangle are at **a,b,c,** the three medians of the triangle meet at $(\mathbf{a}+\mathbf{b}+\mathbf{c})/3$ and this result is independent of where we take the origin.

Alternatively we might define a Cartesian coordinate system with an origin O and three axes OX,OY,OZ, and then specify a point, such as **a**, by giving the components a_x, a_y and a_z. These components clearly depend on where we take the origin, how we choose the axes and also on the unit of length (inches? mm?). If a_x, a_y, a_z, are the components in one coordinate system, then in a new system in which the origin has been displaced by d_x, d_y, d_z, but the axes retain the same directions, the new components will be a_x-d_x, a_y-d_y, a_z-d_z. This is perhaps the simplest of all coordinate transformations and we could express it as

$$x\rightarrow x'=x-d_x, \quad y\rightarrow y'=y-d_y, \quad z\rightarrow z'=z-d_z \tag{2.1}$$

Another kind of transformation would be one in which the axes were rotated leaving the origin unchanged. In this case the new coordinates would be related to the old by a linear transformation of the general form

$$x'=ax+by+cz, \quad y'=dx+ey+fz, \quad z'=gx+hy+iz \tag{2.2}$$

where the constant coefficients a,b,c,d etc. depend on the axis and angle of rotation. Thus, for a rotation of the axes through an angle ϕ about the z axis, we have $a = e = \cos\phi$, $b = -d = \sin\phi$, $i = 1$ with all the other coefficients zero. A further type of transformation occurs when the new system moves relative to the old system with a constant velocity **u** having components u_x, u_y, u_z relative to the original coordinate system. In this case

$$x'=x-u_xt, \quad y'=y-u_yt, \quad z'=z-u_zt. \tag{2.3}$$

This is the classical transformation of Galilean relativity.

It leaves the **form** of all the **classical** dynamic laws unaltered. Thus if a particle of mass m has a velocity $\mathbf{v}$ with components $v_x, v_y, v_z,$ in the first coordinate system and is subject to a force $\mathbf{F}$ so that $mdv_x/dt=F_x$ etc, then in the new system $F'_x=F_x$ and $mdv'_x/dt=F'_x$ etc, Of course this classical transformation would change the velocity of light c to $c-u$ and so could not be compatible with electromagnetism.

2.2 The Lorentz Transformation

Like the Galilean transformation this relates the coordinates of a single event in one inertial coordinate system to the coordinates of the same event in a second inertial coordinate system moving with a uniform velocity relative to the first system, when both systems use the same time standards and length standards. Within each system the clocks are synchronised but now we do not make any assumption that if r is the distance between two points in one system then r', the distance between the same two points referred to the second system, is equal to r, nor that if t is the time between two events in the first system then t' the time between the two events referred to the second system is also equal to t. This, as we shall see, allows us to find a unique transformation that makes the velocity of light the same in all inertial coordinate systems.

Because the only defined direction is that of the relative velocity $\mathbf{u}$ of the two frames of reference we can choose the axes so that OX and OX' coincide in direction with this velocity, and we can also choose the origin of time in both systems to occur when the two origins O and O' coincide. We also choose the OY' and OZ' axes so that, at this instant, they coincide with OY and OZ.

A set of coordinates x, y, z, t given, in the original system S, to a particular event must, at the very least, determine a unique set of coordinates x',y',z',t' in the new system S' and *vice versa*. The **only** transformation with this property of reciprocal uniqueness that does **not** transform **any** neighbouring points in F to points **infinitely** far apart in F' is a **linear** transformation. Thus we must write $x'=\mathrm{A}x+\mathrm{B}y+\mathrm{C}z+\mathrm{D}t, y'=\mathrm{E}x+\mathrm{F}y+\mathrm{G}z+\mathrm{H}t, z'=\mathrm{I}x+\mathrm{J}y+\mathrm{K}z+\mathrm{L}t, t'=\mathrm{M}x+\mathrm{N}y+\mathrm{P}z+\mathrm{Q}t$. [In the theory of optical instruments we can also consider

collinear transformations $x_i' = \Sigma_j\{a_{ij}x_j+b_i\}/\Sigma_k\{f_kx_k+g\}$ where i,j and k run over 1,2,3 corresponding to *x*,*y* and z, and sums over j and k are implied. Here it does not matter if a point where the denominator vanishes is imaged at infinity, as this is just the property of the focus (see Appendix A8).

The successive positions of a particle starting at O and moving along the X axis with a velocity u define a series of events for which $x=ut, y=0, z=0$ in the unprimed system, but the particle is at rest at $x'=0, y'=0, z'=0$ in the primed system. It follows that since x' depends linearly on x and t, it can only do so as $x' = g(x-ut)$ where g may depend on u. Furthermore, since the equations are all linear, should y' and z' depend on t they can only do so with t in the combination $x-ut$. But the line of relative motion is the **only** defined direction in the system and since there are no other preferred directions neither y' nor z' can depend on x. They cannot therefore depend on t, similarly nor can t' depend on either y or z. Because y' does not depend on t an event on the z (or z') axis with $y=y'=0$ at $t=0$ stays on the z' axis and so y' cannot depend on z. Similarly z' cannot depend on y. Finally t' must be related to t and x by $t'=ht+kx$.

A rotation through 180° about the y axis changes x to $-x$, u to $-u$ and leaves t and t' unchanged, thus $h(-u) = h(u)$, $g(-u) = g(u)$ and $k(-u) = -k(u)$. The inverse transformation from the primed to the unprimed coordinates is obviously obtained by changing u to $-u$, for the origin 0 has a velocity $-u$ along the 0'X' axis. Apart from the trivial relation $y'=y$, $z'=z$, we have

$$x' = g(u)[x-ut], \quad t' = h(u)t + k(u)x, \quad (2.6a)$$

$$x = g(u)[x'+ut'], \quad t = h(u)t' - k(u)x'. \quad (2.6b)$$

If Eq.(2.6a) is substituted in Eq.(2.6b) the result is $x = [g^2+ugk]x+u[h-g]t$, $t = [h^2+ukg]t+[hk-gk]x$. These two equations must be identities in x and t and so we have $h=g$, and $k=[1-g^2]/ug$ leading to

$$x' = g[x-ut], \quad t' = g[t+ x(1-g^2)/ug^2], \quad (2.7a)$$

$$x = g[x'+ut'], \quad t = g[t'-x'(1-g^2)/ug^2]. \quad (2.7b)$$

These, together with the simple relations, y'=y, z'=z, are the most general transformation laws (given our choice of the axes and origins) between the coordinates of an event in two different **inertial** coordinate systems with a relative velocity u. They have arisen solely because we have assumed (i) that space and time are homogeneous, i.e. there are no preferred positions or instants, (ii) that space is also isotropic without any preferred directions and (iii) that if Newton's first law holds in one coordinate system it also holds in any other system in uniform relative motion. Apart from u they depend on one further parameter g which may be a function of u, though not, of course, of x,y,z or t. If we wish to retain the intuitive notion that $t'=t$, then we must put $g=1$ and this reduces Eqs.(2.7a,b) to the Galilean transformation of classical mechanics i.e.

$$x' = x - ut, \quad y' = y, \quad z' = z\ , \quad t' = t. \tag{2.8}$$

So far, however, the function $g(u)$ is arbitrary, except that it must be an even function of u and reduce to unity as u tends to zero.

We can now impose a further condition, the invariance of the velocity of light c. Let $x=vt$, then $x'=g(u)(v-u)t$ and $t'=g(u)[1+v(1-g^2)/ug^2]t$, giving $x'=[v-u]t'/[1+v(1-g^2)/ug^2]$. We now require that if $v=c$, then $v'\equiv x'/t'=c$. This leads to $c[1+c(1-g^2)/ug^2]=c-u$ giving the relation $g^2=1/(1-u^2/c^2)$. Since g must tend to +1 as u tends to zero we finally have

$$\gamma(u)\equiv g(u) = (1-u^2/c^2)^{-\frac{1}{2}}\ , \tag{2.9}$$

where we have replaced g by the traditional Greek letter γ to give a formula with which we shall soon become very familiar.

The complete Lorentz transformation therefore becomes

$$x'= \gamma(u)[x-ut] = (1-u^2/c^2)^{-\frac{1}{2}}[x-ut], \tag{2.10a}$$

$$y' = y, \qquad z' = z, \tag{2.10b}$$

$$t'= \gamma(u)[t-ux/c^2] = (1-u^2/c^2)^{-\frac{1}{2}}[t-ux/c^2], \tag{2.10c}$$

and the inverse transformation is

$$x = \gamma(u)[x'+ ut'], \tag{2.11a}$$

$$y = y', \quad z = z', \tag{2.11b}$$

$$t = \gamma(u)[t'+ux'/c^2]. \tag{2.11c}$$

These two sets of equations relate the coordinates and times (x,y,z,t) and (x',y',z',t') given to the **same** identical single event in two coordinate systems which are defined in two different inertial frames of reference, such that the origin of the primed frame moves along the OX axis of the unprimed frame with a constant velocity u, and the origins and axes of the two frames coincide at $t=t'=0$. They describe the *only* transformation that incorporates the isotropy of space (there are no preferred directions) and the uniformity of space and time (no preferred locations or instants), that preserves Newton's first law and, finally, that leaves the velocity of light c as a universal constant, the same in all inertial frames of reference.

2.3 *Simple Kinematic Results*

Because *c*, initially introduced as the velocity of light, has now appeared in equations describing space-time relations it will appear in other parts of physics having nothing to do with electromagnetism. In particular, as we shall see in a later chapter, if we wish to retain the law of conservation of linear momentum when speeds comparable with *c* are involved, the usual classical relation $\mathbf{p}=m\mathbf{v}$ between momentum $\mathbf{p}$, mass *m* and velocity $\mathbf{v}$ has to be replaced by

$$\mathbf{p} = m\mathbf{v}/(1-v^2/c^2)^{\frac{1}{2}} = m\mathbf{v}\gamma(v),$$

and this has profound repercussions throughout physics.

One way of looking at this relation between momentum and velocity is to suppose that the mass *m* of a particle of velocity *v* is related to its **rest mass** m_0 by $m = m_0\gamma(v)$, and Kaufmann (1902), three years before Einstein's paper and a year before Lorentz published his transformation, had (in a version of J.J.Thomson's experiments) observed just this variation with velocity of the mass-to-charge ratio of an electron, but had ascribed it to an electromagnetic effect!

We look first at some purely kinematic consequences of the Lorentz transformation. Obviously the most surprising aspect of the transformation is that t and t', the times ascribed to the same event, are different in the two coordinate systems. This, alone, might be enough to make us have doubts about the relevance of the transformation to real physics, were it not for the way that nature provides a simple, accurate, direct and immediately comprehensible experimental demonstration of the time transformation.

Muons, particles first detected in cosmic rays, decay when they are at rest (or only moving slowly), about 2 microseconds after their formation through the decay of a pion. It is, however, also possible to observe the decay of muons with velocities approaching *c*. Now suppose that a muon is formed at rest in a frame F' with a velocity u relative to the laboratory frame of reference F in which the coordinates of an event are x,y,z,t (henceforth we use "coordinates" to include time as well as space). We take it, for convenience, to be formed at $t'=t=0$, and to decay at $t'=\tau$ in its own rest frame F'. Thus τ is its lifetime at rest. It also decays at the same place x' (in its rest frame) at which it was formed and therefore we might as well take $x'=0$. In the laboratory frame F it is formed at $x=0$ and decays, according to Eq. (2.11a), at $x=\gamma(u)ut'=\gamma(u)u\tau$ and, according to Eq. (2.11c), at $t=\gamma(u)\tau=\tau/(1-u^2/c^2)^{1/2}$. This time t can be considerably greater than τ. For example if the muons are created with a velocity 0.99c the observed life time will be some 7 times longer than the natural lifetime τ. Experiments with muons trapped in a storage ring have verified the relation

$$t = \tau/(1 - u^2/c^2)^{1/2} \tag{2.12}$$

up to values of t/τ greater than 100. This effect, **"time dilation"**, is what Poincaré meant by saying that moving clocks must run slow.

In the laboratory frame F, between its formation and its decay, the muon moves a distance $ut = u\tau/(1-u^2/c^2)^{1/2}$ which, if u is comparable to c, can be much greater than τc. In fact muons formed in the upper atmosphere at heights up to 10 km reach the earth's surface. Even travelling with the velocity of light this would take 30 microseconds, much longer than the natural lifetime of the muon. Of course this result is also one of the staples of science fiction for it allows voyagers with a natural life-time of only three score years and ten to travel to galaxies many, many thousands of light years away (if, that is, they could be accelerated to very nearly the velocity of light!).

A second simple kinematic result is the formula for combining velocities. The trajectory of a particle travelling along the 0'X' axis of the frame F' with a velocity v' is given by $x'=v't'$. Its description in the laboratory frame F is $x = \gamma(u)[x'+ut'] = \gamma(u)[v'+u]t'$ and $t = \gamma(u)[t'+ux'/c^2] = \gamma(u)[1+uv'/c^2]t'$ so that in the laboratory frame the velocity is

$$v = x/t = [v'+ u]/[1 + uv'/c^2]. \tag{2.13}$$

We note that if either u or v' is equal to c, then $v = c$.

Combining any velocity with c always gives c.

We can use Eq.(2.13) to calculate $\gamma(v)$ in terms of u and v'. We begin with $1-v^2/c^2 = [(1+uv'/c^2)-(u+v')^2/c^2]/[1-uv'/c^2]^2$ which simplifies to give

$$\gamma(v) = \gamma(u)\gamma(v')[1 + uv'/c^2]. \qquad (2.14)$$

This is itself a useful result with many applications but for the moment we note that, whatever their signs, as long as u and v' are numerically less than c, the right hand side of Eq.(2.14) is positive, finite and real, thus $\gamma(v)$ is also positive, finite and real so that v must be numerically less than c.

Combining two velocities less than c can never yield a velocity greater than, or even equal to c.

Because times and coordinates must be expressed by real numbers, the Lorentz transformation can only be a proper physical relation if $\gamma(u)=1/(1-u^2/c^2)^{1/2}$ is real and therefore if $u \le c$. It is impossible to construct meaningful relations between frames of reference whose relative velocity exceeds c. On the other hand if a body has a velocity u relative to another body, we could certainly give a meaning to the two frames in which one or the other body is at rest. We must conclude that when $u>c$ there can be no relation between events associated with both the bodies. This can be expressed by saying that c is the greatest velocity that can appear in any equation expressing an observable relation between any two bodies. Thus not only is c the invariant velocity of light but it is also the greatest velocity with which **signals of any kind** can propagate. Any theory that ignores this leads to nonsensical paradoxes in dealing with causality. This does not mean that velocities greater than c cannot occur. As a trivial example if a pair of scissors is closed the point where the blades intersect moves forward much faster than the blades move together, and there is nothing in relativity to say that the point of intersection cannot move faster than c. As Trester (1989) points out the distinction is between the propagation of **effects** and of **signals**. It is also worth remarking that although in most cases waves propagate signals at their group velocity, this is not a universal rule. The subtle distinction between group velocity and signal velocity is discussed in detail by Brillouin (1960).

Whereas time dilation is readily observed experimentally the corresponding spatial effect, called the Lorentz contraction, is not. In general, objects large and solid enough to have a well defined length at rest are too massive to be accelerated to anywhere near the velocity of light. Nevertheless an appreciation of the effect is often helpful in understanding the content of a relativistic calculation. Consider a rod which, when at rest (in its rest frame), has a length L' and which is moving along the 0X axis parallel to its length with a velocity u. It is illuminated by a single flash of light at $t=0$ when the trailing end of the rod is at $x = 0$ and the leading end is at $x = L$. In the rod's rest frame this is at $x'=L'$, but $x'=\gamma(u)[x-ut]$ and, since $t=0$, this gives $L'=\gamma(u)L$ or

$$L = L'(1 - u^2/c^2)^{1/2} , \tag{2.15}$$

and the length L in the laboratory F frame is **less** than the length L' in the rod's rest frame F'. Although both ends of the rod are seen in the frame F at the same time, they are seen at different times in the frame F' moving with the rod. Suppose instead that the rod were to be illuminated all at the same time $t' = 0$ in the frame F'. Then the leading edge of the rod would appear at $x = \gamma(u)[x'+ ut'] = \gamma(u)L'$ giving

$$L = L'/(1 - u^2/c^2)^{1/2} \tag{2.16}$$

and the rod would appear **longer**, but now of course the two ends are seen in the laboratory frame F at different times $t=0$ and $t=\gamma(u)uL'/c^2=uL/c^2$, and between these two times the rod has moved a distance u^2L/c^2. As we remarked earlier, the Lorentz contraction of a macroscopic body is not a directly observable experimental effect: the energy needed to accelerate it to anything like the velocity of light is prohibitive.

The Lorentz transformation only relates the coordinates to be ascribed to events in two different frames of reference in relative motion, it does not tell us directly how an object specified in one frame, for example as a solid cube, might appear in the other frame. Discussions of the appearance of an object moving with nearly the velocity of light have very little relevance to the world of observable physical phenomena.

2.4 General Axes

It may not always be convenient to take the OX and O'X' axes along the direction of the relative motion of the two frames of reference F and F' in which the coordinate systems *C* and *C'* are defined. The results can be generalised by first rotating the axes in *C* to the standard position with OX along the motion, then making the Lorentz transformation, and finally rotating the axes in *C'* back to their proper position. Here we try another seemingly reasonable appproach and show why it fails.

There appears to be no obvious reason why the axes in *C'* and *C* should not be parallel to each other even if their relative velocity *u* is not along one of the axes. Let us see where this leads. Rather than letting *u* have an arbitrary direction relative to the axes, we continue to take OX and O'X' along the line of motion and define the other axes as before, but then consider two new sets of Cartesian axes OP, OQ, OR, and OP' OQ' OR'. Ignoring the third dimension the situation can be illustrated as in Figure 2.1. Now if OP' is parallel to OP and OQ' to OQ then at least the axes coincide at $t = t' = 0$ when the two origins coincide. Consider a point p on OP with components x_p, y_p, and a point q on OQ with components x_q, y_q, relative to the standard axes. In the frame F', and using the x',y' axes, we have when $t=0$, $x_p'=\gamma(u)x_p$ and $y_p'=y_p$ and also $x_q'=\gamma(u)x_q$ and $y_q'=y_q$

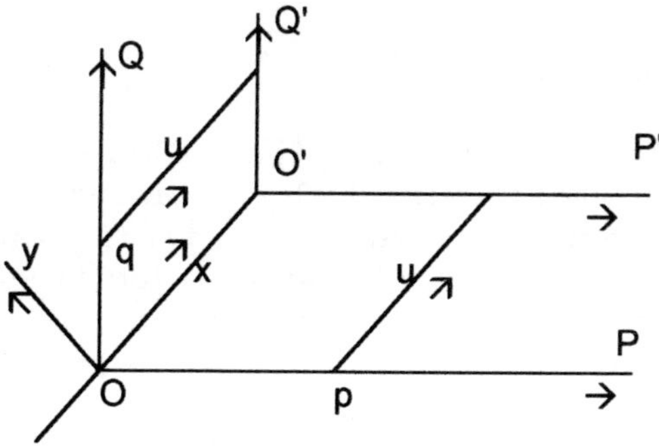

Figure 2.1 General axes

In F, OP and OQ are orthogonal so that the scalar product $x_p x_q + y_p y_q = 0$, but in F' we have $x_p'x_q'+y_p'y_q'= \gamma^2 x_p x_q + y_p y_q = (\gamma^2-1)x_p x_q$, and this is not zero unless either $u=0$, or $x_p=0$ or $x_q=0$, i.e. unless OP or OQ coincide with the standard axes. Thus only the standard axes with one axis along the line of motion can be supposed to be both orthogonal, or Cartesian, in each frame and also parallel at t=0. Clearly the notion of parallelism in two frames in relative motion needs handling with care if it is not to lead to paradoxical and nonsensical results.

We now look at an alternative and, in this case, correct approach. Let the vector **r** be the location of an event at time t in the frame of reference F, and let F' have a vector velocity **u** relative to F, then, in terms of the scalar product **r.u**, the component of **r** parallel to **u** is

$$\mathbf{r}_{/\!/} = \mathbf{u}(\mathbf{r}.\mathbf{u})/u^2 \,, \qquad (2.17a)$$

and the component perpendicular to **u** is $\mathbf{r} - \mathbf{r}_{/\!/}$. We can now use the Lorentz transformation in its standard form to obtain the components of **r'** in F'.

$$\mathbf{r}_{/\!/}' = \gamma(u)[\mathbf{r}_{/\!/} - \mathbf{u}t], \qquad \mathbf{r}' - \mathbf{r}_{/\!/}' = \mathbf{r} - \mathbf{r}_{/\!/}, \qquad (2.17b)$$

and this leads to

$$\mathbf{r}' = \mathbf{r} + \mathbf{u}[\{\gamma(u)-1\}\mathbf{u}.\mathbf{r}/u^2 - \gamma(u)t] \,, \qquad (2.17c)$$

$$t' = \gamma(u)[t - \mathbf{u}.\mathbf{r}/c^2]. \qquad (2.17d)$$

These useful relations work because **r** and **r'** have been separated into their components parallel and perpendicular to **u**, the vector velocity specifying the only defined direction in the relation between the reference frames.

2.5 Relative Velocity

We will often have to find the relative velocity of two particles given their individual velocities **u** and **v** referred to a common inertial frame F. We first transform to the frame in which the particle with velocity **u** is at rest. This is easiest if the OX and OX' axes are parallel to **u** .

A displacement dx,dy,dz and a time dt in F become, in F', $dx'=\gamma(u)(dx-u\,dt)$, $dy'=dy$, $dz'=dz$, $dt'=\gamma(u)(dt-u\,dx/c^2)$. For the particle of velocity **v** we have d**r**=**v**dt and so

$$v_x' = (v_x-u)(1-uv_x/c^2)^{-1}, \qquad (2.18a)$$

$$v_y' = v_y[\gamma(u)(1-uv_x/c^2)]^{-1}, \qquad (2.18b)$$

$$v_z' = v_z[\gamma(u)(1-uv_x/c^2)]^{-1}. \qquad (2.18c)$$

Notice how the transverse components involve both $\gamma(u)$ and $(1-uv_x/c^2)$. In vector notation these expressions become

$$\mathbf{v}_{\parallel}' = \mathbf{u}(\mathbf{v}.\mathbf{u}-u^2)/[u^2(1-\mathbf{u}.\mathbf{v}/c^2)], \quad (2.19a)$$

$$\mathbf{v}_{\perp}' = \mathbf{v}_{\perp}/[\gamma(u)(1-\mathbf{u}.\mathbf{v}/c^2)], \quad (2.19b)$$

and these can be combined to give

$$\mathbf{v}' = [\mathbf{u}\{\gamma(u)-1\}\mathbf{v}.\mathbf{u}/u^2+\mathbf{v}-\mathbf{u}\gamma(u)]/[\gamma(u)(1-\mathbf{u}.\mathbf{v}/c^2)]. \quad (2.19c)$$

The square of the magnitude of $\mathbf{v}'$ is

$$|v'|^2=[(v_x-u)^2+(1-u^2/c^2)(v_x{}^2+v_y{}^2)]/(1-\mathbf{u}.\mathbf{v}/c^2)^2 \quad (2.20a)$$

which can also be expressed as

$$|v'|^2=c^2[1-(1-u^2/c^2)(1-v^2/c^2)(1-\mathbf{u}.\mathbf{v}/c^2)^{-2}] \quad (2.20b)$$

and from this we obtain the result

$$\gamma(v') = \gamma(u)\gamma(v)[1-\mathbf{u}.\mathbf{v}/c^2]. \quad (2.21)$$

If the velocity $\mathbf{v}'$ of a particle is given relative to a particle that itself has a velocity $\mathbf{u}$ relative to F, then the first particle has a velocity

$$\mathbf{v} = [\mathbf{u}\{\gamma(u)-1\}\mathbf{u}.\mathbf{v}'/u^2+\mathbf{v}'+\mathbf{u}\gamma(u)]/[\gamma(u)(1+\mathbf{u}.\mathbf{v}'/c^2)] \quad (2.22)$$

and this leads to

$$\gamma(v) = \gamma(v')\gamma(u)(1+\mathbf{u}.\mathbf{v}'/c^2). \quad (2.23)$$

Eqs.(2.21) and (2.23) are especially important for, as we shall see later, the energy of a particle of velocity v and rest mass m is $mc^2\gamma(v)$.

2.6 Successive Lorentz Transformations

Relations between the coordinate systems in three frames F, F' and F" will only be simple if the velocity $\mathbf{v}'$ of F" relative to F' is parallel to the velocity $\mathbf{u}$ of F' relative to F. Consider the case where F' has a velocity u along the x axis of F and F" has a velocity v' along the y' axis of F'. Clearly we can properly assert that the axes in F and F' coincide at $t = t' = 0$ and also that the axes in F" and F' coincide at the same time $t''=t'$. This allows us to calculate $\mathbf{V}$, the velocity relative to F of the origin O" of F", and its components are $V_x = u$, $V_y = v'/\gamma(u)$. We can also calculate $\mathbf{v}''$, the velocity of O, the origin of F, relative to F", and this has components $v_x''=-u/\gamma(v')$ and $v_y''=-v'$. Thus in F the tangent V_y/V_x of the angle θ between the line of this relative motion and the x axis is $v'/[u\gamma(u)]$ while in F" the tangent of the angle θ'' between the motion and the x" axis is $v_y''/v_x'' = v'\gamma(v')/u$. We see that these two angles are unequal and that θ'' is always larger than θ. Since the direction of the relative motion of the origins has a real physical significance we can only conclude that although the axes in F and F" may each be parallel to the axes in F', the axes in F" cannot be parallel to the axes in F. Results such as this should warn us to beware of appeals to intuition in discussing relativistic problems.

If, although u is finite, $v' \to dv'$ is infinitesimal, so that $\gamma(v') \to 1$ and the angles θ and θ'' are small, the situation is simpler and gives $d\theta^\star \equiv \theta''-\theta = [\gamma(u)-1]dv'/[u\gamma(u)]$. The angle $d\theta^\star$ is defined in F, and in this frame $-d\theta^\star$ is the angle between the x" and the x axes. Although u is also defined in F, dv' is defined in F'. In F it corresponds to $dv = dV_y$ and this is equal to $dv'/\gamma(u)$. Thus we have $d\theta^\star = [\gamma(u)-1]dV_y/u$ expressed entirely in terms of velocities relative to F. It describes a rotation (the Thomas Precession) of the axes through an angle proportional to the component of $d\mathbf{v}$ perpendicular to $\mathbf{u}$ and about an axis perpendicular to both $d\mathbf{v}$ and $\mathbf{u}$. It can be expressed in vector notation, using the vector product $\times$, as

$$d\theta = \theta - \theta^* = -d\theta^\star = [\gamma(u)-1]d\mathbf{v}\times\mathbf{u}/u^2 \qquad (2.24a)$$

Since $u^2 = c^2[1-\gamma^{-2}]$ this gives

$$d\theta = \gamma^2(u)d\mathbf{v}\times\mathbf{u}/[c^2(\gamma(u)+1)] \qquad (2.24b)$$

and, if u is appreciably less than c, so that $\gamma(u)\approx 1$, then

$$d\theta \approx d\mathbf{v}\times\mathbf{u}/2c^2. \qquad (2.24c)$$

We will give an application of this result in the next chapter.

2.7 The Invariant Interval

If $\mathbf{R}$ and $\mathbf{r}$ are the positions of two events in the frame F which occur at times T and t, and their Cartesian components are R_i and r_i with i=1,2,3, then, when the axes in F are rotated the individual components with respect to the new axes change to $R_i^\star$ and $r_i^\star$, but sums such as $\sum_i(R_i-r_i)^2$ i.e. $|\mathbf{R}-\mathbf{r}|^2$, the square of their spatial separation, are unchanged and, of course, $T^\star-t^\star=T-t$. However a transformation to a frame F' with a velocity u along the x axis of F yields

$$(R_i'-r_i')^2=(R_2-r_2)^2+(R_3-r_3)^2+\gamma^2[R_1-r_1-u(T-t)]^2,$$
$$(T'-t')^2 = \gamma^2(u)[T-t-u(R_i-r_i)/c^2]^2, \qquad (2.25)$$

and neither the time difference nor the spatial separation

is separately invariant, but consider the quantity

$$\tau^2 = c^2(T-t)^2-(R_i-r_i)^2 \qquad (2.26)$$

which transforms to $\tau'^2=c^2(T'-t')^2-(R_i'-r_i')^2$.
Direct substitution from Eq.(2.24) yields

$$\tau'^2 = \tau^2 \qquad (2.27)$$

and so τ is an invariant, known as the **invariant interval.**

With $T=t+\mathrm{d}t$ and $\mathbf{R}=\mathbf{r}+\mathrm{d}\mathbf{r}$ we can also define the infinitesimal invariant interval

$$\mathrm{d}\tau = [c^2(\mathrm{d}t)^2- (\mathrm{d}r)^2]^{1/2}. \qquad (2.28)$$

As this only refers to an infinitesimal region of space-time it can have a **local** significance in frames which as a whole are not inertial, much as the instantaneous velocity of a particle has a meaning even if the particle is accelerating. It could for example be used to obtain a (time-dependent) transformation relating the coordinates ascribed to events in an accelerating frame (e.g. one fixed in a rocket) and the coordinates of the same events in an inertial frame.

2.8 Past and Future: The Light Cone

An event B following an event A after a time t and at a distance r could only have been influenced by A if $r \leq ct$. If the interval $\sigma^2 = c^2t^2-r^2$ is positive it is called time-like and if negative it is space-like. Events with a time-like separation can, but events with a space-like separation cannot have a causal connection. If the separation between B and A is time-like and t is positive then B is part of A's future, if t is negative B is part of A's past.

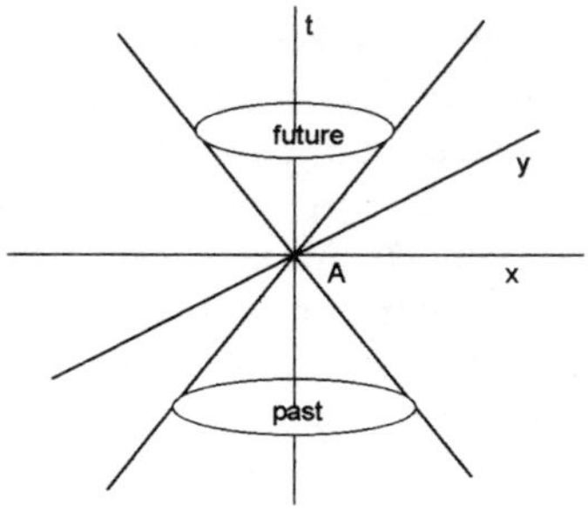

Figure 2.2 The Light Cone

By confining attention to only two space dimensions these relations can be illustrated by a diagram (Figure 2.2) showing the *light-cone*, $\sigma^2 = 0$ which divides the space-time associated with an event A at the origin into three distinct regions, the past of A, the future of A and the rest of space-time with a space-like separation from A. This region contains only those events with no causal connection with the event A. If A describes the present position of a particle, its past position must have been somewhere inside the past light-cone and its next position will be within the future cone. If the particle is at rest its "trajectory" lies upwards along the vertical time axis. Since the interval is invariant, the identification of a space-time region containing the entire possible past and future related to a particular event is invariant, and so is the separation between past and future; for $t'=\gamma(u)[t-ux/c^2]$ and so, if $\sigma^2 = c^2t^2-x^2$ is positive (to give a time-like separation) we certainly have $(ux/c^2)^2<t^2$ so that t and t' must have the same sign. Thus relativity at least leaves intact this very fundamental intuition about time. It should not, however, be assumed that a particle or a space traveller at A with a finite life-time T cannot reach regions where $r>cT$. If the particle has a velocity u in the frame F (to which the light-cone refers) the appropriate life-time to use is not T, but $T^* = \gamma(u)T = T/(1-u^2/c^2)^{1/2}$. Thus if, for example, u=0.99c the particle can reach regions as far away as r=7cT.

2.9 Proper Time

Events occurring **within** a body moving through the laboratory frame F with a velocity u are most naturally discussed in terms of the rest frame of the body, i.e. the frame which moves with the body through the laboratory frame. Two events that occur a time T apart, but at the same place in the body, occur at different places, as well as a different time t apart, in the laboratory frame, but T, rather than t, is clearly the right time to use in discussing events and processes that occur within the body. It is called the **proper time**. The proper time between events is the time in a frame in which the system wherein the events occur is at rest. It is clearly independent of any choice of a frame in which to describe the system's motion; for example, the proper life-time of a muon is always the same and equal to the *measured* life-time for muons at rest in the laboratory frame. Because of this invariant significance the proper time plays an important role in formulating relativistic mechanics. The proper time associated with a moving body is the ordinary time in the *comoving frame*, a frame moving with the body. Apart from a factor *c*, the proper time is the invariant interval between events occurring at the same place within a body.

2.10 Discussion

The Lorentz transformation is the central mathematical result of special relativity and it alters some of our basic ideas about space, time and motion. In the next chapter we will look at a selection of kinematic effects i.e. effects that do not involve any discussion of interactions (forces) between moving bodies. Not until Chapters 4 and 5 will we begin to consider dynamics, and there we shall see how the Lorentz transformation alters this most fundamental branch of physics. Since classical dynamics is not compatible with the Lorentz transformation, the classical relations between mass, velocity, momentum, force, work and energy have to be modified. These modifications affect almost every branch of physics even if only indirectly.

Problems

2.1 F and F' are two frames of reference with a common fixed origin and F' rotates relative to F with a constant angular velocity $\boldsymbol{\omega}$. Show that if $\mathbf{S}'$ is a vector quantity defined relative to F' and $\mathbf{S}$ is the same quantity referred to F then $d\mathbf{S}/dt = d\mathbf{S}'/dt + \boldsymbol{\omega}\times\mathbf{S}$. Let $\mathbf{S}'$ be the position relative to F' of a particle of mass m which moves relative to F' with a constant velocity $\mathbf{v} = d\mathbf{S}'/dt$. Find its acceleration relative to F.

2.2 Two events occur at $t=0$ and at $x=0$ and $x=10$km in F. How far apart in time and space do they occur in a frame F' moving along the x axis of F with a velocity $u=10^5$km/sec?

2.3 A rocket flies at 3000m/sec out to Uranus, which is 3.10^9km away. How long does the flight last? What will be the difference between this time and the flight time recorded by a clock on the rocket?

2.4 A particle emitted in an energetic collision between two nucleons has a lifetime of 10^{-18}sec at rest. How far does it travel before it decays if it is emitted with a velocity a) 0.9c, b) 0.99c, c) o.999c, d) 0.9999c?

2.5 Two particles approach each other along a straight line with equal velocities 2.10^8 m/sec. Find their relative velocity.

2.6 Two particles travel at velocities $0.9999c$ and $0.9c$ in the same direction. Find their relative velocity.

2.7 The frame F' has a velocity $u=\frac{1}{2}c$ along the X axis of F. Find the components of the velocity in F' of a particle whose velocity relative to F has components $(\frac{1}{2}c, \frac{1}{2}c, 0)$.

2.8 If the vector velocity of F' is $\mathbf{u}=(\frac{1}{2}c, \frac{1}{2}c, 0)$ and the velocity of a particle in F is $\mathbf{v}=(\frac{1}{2}c, \frac{1}{2}c, 0)$, find the magnitude of the velocity $\mathbf{v}'$ in F'.

2.9 A body travelling along the 0X axis of a coordinate system C with a velocity u breaks up into two equal fragments whose motions are parallel to 0X and whose relative velocity is v. Find the velocities of the two fragments relative to C if $u=v=2c/3$.

2.10 Show that if $\mathbf{u}$ increases by $d\mathbf{u}$, $\gamma(u)=(1-u^2/c^2)^{-\frac{1}{2}}$ increases by $\gamma^3(u)\mathbf{u}.d\mathbf{u}/c^2$.

CHAPTER 3

KINEMATIC AND OPTICAL EFFECTS

In the last chapter we looked at three basic kinematic effects: time dilation, the Lorentz contraction and the relativistic rules for combining velocities. We now look at some slightly more complicated effects.

3.1 The Doppler Effect

We consider an atom which, in its rest frame, emits electromagnetic waves of frequency f'. The atom moves with a velocity $\mathbf{u}$ and we ask what frequency will be observed by an observer at a fixed point in the laboratory frame F. We begin with the case where the atom is moving directly towards the observer. During an interval t measured in the laboratory frame the atom has moved a distance ut towards the observer and so waves emitted in a time t reach the observer in a time $t^*=t(1-u/c)$. The interval t in the laboratory frame corresponds to an interval $t'=t(1-u^2/c^2)^{1/2}$ $=t/\gamma(u)$, in the rest frame of the atom. In this interval it emits $n=f't'$ waves. These n waves are then received by the observer during $t^* = t(1-u/c) = t'(1-u/c)/(1-u^2/c^2)^{1/2} = t'\{(1-u/c)/(1+u/c)\}^{1/2}$ and so the observed frequency is

$$f=n/t^*=f'[(1+u/c)/1-u/c)]^{1/2} \qquad (3.1a)$$

If the atom is moving away from the observer the corresponding result is obviously

$$f=f'[(1-u/c)/(1+u/c)]^{1/2}. \qquad (3.1b)$$

When the atom is moving along a line normal to the line to the observer there is no true Doppler shift, all that remains is time dilation and so

$$f=f'[1-u^2/c^2]^{1/2} \qquad (3.1c)$$

This is often called the transverse Doppler shift. Unlike the longitudinal effects, linear in u/c, the transverse effect is quadratic in u/c .

In a solid with a lattice thermal capacity C and an atomic weight M the fractional change in frequency due to lattice vibrations is $d\nu/\nu = -C/2Mc^2$ deg^{-1} and for iron at room temperature this gives -2.21×10^{-15}deg^{-1}. This has been verified by Pound and Rebka (1960) to about 5% accuracy using the Mossbauer effect in Fe^{57} (see #7.7).

The classical Doppler shift is generally associated with sound. There is no transverse effect and, because the waves now propagate in a definite medium, there is a difference between the case when the observer moves relative to the medium and the case when the source moves.

Consider first the case when the observer moves. At $t=0$ let the observer be a distance ℓ from the source, then the wave arriving at this time left the source at $t_1=-\ell/s$, where s is the velocity of sound. At time t the distance between the source and the observer is reduced to $\ell-ut$ and the wave arriving at t left the source at $t_2=t-(\ell-ut)/s$. Between t_1 and t_2 the source emitted $f'(t_2-t_1)$ waves and these arrived at the observer in a time t, thus the perceived frequency is

$$f=f'(t_2-t_1)/t=f'(1+u/s). \qquad (3.2a)$$

When the source moves, a wave reaching the observer at $t=0$ again left the source at $t_1=-\ell/s$ but a wave reaching the observer at t left the source at $t_2=t-[\ell/s-(t_2-t_1)u/s]=t+t_1+(t_2-t_1)u/s$, so that $t=(t_2-t_1)(1-u/s)$, and the perceived frequency is

$$f=f'/(1-u/s). \qquad (3.2b)$$

To first order in u/s (3.2b) is the same as (3.2a) and also, if the velocity of sound is replaced by c, as the relativistic formula Eq.(3.1a).

Because there is no classical transverse effect, it might seem easier to search for a relativistic effect, although it is quadratic in u/c, by observing the transverse effect. At the time of the first experiments it was difficult, even with electrically accelerated ions to achieve values of $u^2/2c^2$ much greater than 10^{-4} and so if this small quadratic effect was not to be confused with the much larger linear longitudinal effect it would have been necessary to keep the line of sight perpendicular to the motion to within about 10^{-2} radians or half a degree. The required degree of collimation of both the beam of ions and the optical path results in too low an intensity for a measurable effect. However, the classical formula gives the average of the wavelengths for a source moving towards and away from an observer as $\tfrac{1}{2}(\lambda_t+\lambda_a)=\tfrac{1}{2}\lambda(1+u/c+1-u/c)=\lambda$ whereas the relativistic result is $\tfrac{1}{2}\lambda[(1-u/c)^{1/2}/(1+u/c)^{1/2} + (1+u/c)^{1/2}/(1-u/c)^{1/2}]$

which gives $\lambda(1 + u^2/2c^2)$ and thus a small quadratic effect, equal to the transverse effect but, as it requires much less stringent collimation, easier to observe experimentally. This was the approach adopted by Ives and Stilwell (1938), who observed the longitudinal effect in the emission spectrum from H_2^+ ions with velocities up to $c/200$, and, although the effect was small, they found complete agreement with the relativistic formula. More recently Alvager et al.(1964) have found much larger effects for the γ rays emitted in the decay of the neutral pion. Further strong evidence is provided by the results obtained with the free electron laser (see #3.2).

It will be apparent that there are two separate factors that contribute to the Doppler effect. The first is the purely relativistic time dilation effect that is quadratic in v/c and entirely responsible for the transverse effect, and then there is the essentially geometric effect, linear in v/c that dominates the longitudinal Doppler effect.

In a wave guide, the frequency ν and the wave number κ are related by the dispersion relation $\nu^2 = \alpha^2 + c^2\kappa^2$ where α is a constant depending on the cross-section of the guide, and the product of the group and phase velocities is c^2. At the cut-off frequency $\nu=\alpha$, the phase velocity $c' = \nu/\kappa$ is infinite, though the group velocity $d\nu/d\kappa$ is zero. If, then, an atom moves with a velocity v along the axis of a waveguide and radiates at exactly the cut-off frequency of the guide, the geometrical contribution to the Doppler effect which is proportional to v/c' will vanish, though the quadratic time dilation effect which arises from the basic structure of space time will be unaffected by the properties of the waveguide.

This effect occurred in the first quantum mechanical amplifier, the ammonia maser, actually an oscillator working at a wavelength near 12mm, (Gordon, Ziegler and Townes 1955), and eliminated the longitudinal Doppler effect due to the thermal spread in the velocities in the molecular beam of ammonia.

For a beam of molecules of mass M at a temperature T, the quadratic effect gives a frequency spread of order kT/Mc^2, which for ammonia at room temperature is about 1 part in

10^{12} whereas the linear longitudinal effect is about the square root of this and thus a part in 10^6.

3.2 The Free Electron Laser

If an electron of velocity u moves past a spatially periodic magnetic structure then, in the electron's rest frame, the spatial period λ of the structure is Lorentz contracted to $\lambda'=\lambda(1-u^2/c^2)^{\frac{1}{2}}$. Alternatively, since the time in the laboratory frame between an electron passing successive periods of the magnetic field is $t = \lambda/u$, we get, for the corresponding time in the electron's rest frame,

$$t'=(t-u\lambda/c^2)/(1-u^2/c^2)^{\frac{1}{2}}=(1-u^2/c^2)^{\frac{1}{2}}\,\lambda/u.$$

The varying magnetic field causes the electron to accelerate and decelerate and therefore radiate at a frequency $f'=1/t'$ (in the rest frame of the electron). An observer in the path of the oncoming electron will see this Doppler shifted to a higher frequency

$$f = f'(1+u/c)^{\frac{1}{2}}/(1-u/c)^{\frac{1}{2}} = u/[\lambda(1-u/c)].$$

In terms of the wave length λ'' of the radiation we obtain

$$\lambda''/\lambda = (c/u)-1 = \gamma/(\gamma^2-1)^{\frac{1}{2}} -1 \ . \qquad (3.3a)$$

When u is nearly equal to c and $\gamma(u)$ is large this gives

$$\lambda''/\lambda \approx 1/(2\gamma^2), \qquad (3.3b)$$

and, anticipating the results in the next chapter, we can express this in terms of the electron energy E and its rest energy mc^2 (about ½ MeV).

$$\lambda''/\lambda = \tfrac{1}{2}(mc^2/E)^2 = \tfrac{1}{2}[1/(1+T/mc^2)]^2, \qquad (3.3c)$$

where T is the kinetic energy of the electron.

As an example electrons accelerated through 50 MeV give $\lambda''/\lambda = 4{\cdot}9.10^{-5}$ and so, with a magnetic structure of period 10mm, the wave-length of the radiation would be at 490 nm in the visible region of the spectrum.

The first free electron laser dates back to the experimental and theoretical studies of Motz (Motz 1951) and (Motz et al.1953) who first demonstrated the generation of visible light using a 100 MeV beam at the Stanford Linac laboratory. Later developments due to Madey (1971) revived interest in the free electron laser as an easily tuneable device, potentially capable of generating substantial power at frequencies in the electromagnetic spectrum from the far infra-red up to the X-ray region. For a fuller account see

"Undulators and Free Electron Lasers" (Luchini and Motz 1990).

3.3 The Sagnac Effect

Though comparisons of times at different places when motion is involved clearly invite a relativistic treatment, this should not lead us to ignore the possible coexistence of classical effects which may be larger. The Doppler shift is a case in point. Very often these classical effects will be linear in the ratio u/c while the relativistic effects will only be quadratic. The effect first discussed by Sagnac in 1915 occurs when light is guided round a closed path by a rotating system of mirrors or a ring of optical fibre. It influences certain types of interferometer, optical gyroscopes, the ring laser, and the comparison of terrestial clocks at different longitudes.

We consider a system of mirrors rotating with a constant angular velocity $\mathbf{\Omega}$ about an axis fixed in an inertial frame of reference, and we let $\boldsymbol{r}$ be the radius vector in the rotating frame from a fixed point on the axis to a point on the light path. We assume that $\Omega r \ll c$ so that we can ignore all relativistic effects (which will be at most of order $(\Omega r/c)^2$). If light passes the point specified by $\boldsymbol{r}$ and proceeds along an element $d\ell$ measured in the rotating frame then, by the time dt that it reaches the end of $d\ell$, this end will have moved through $\mathbf{\Omega}\times\mathbf{r}dt$ in the inertial frame. The corresponding displacement is $d\mathbf{L} = \mathbf{\Omega}\times\mathbf{r}dt + d\ell$ and, to first order, $dL - d\ell = (\mathbf{\Omega}\times\mathbf{r}dt).d\ell/d\ell$. To this order the effect of rotation is to increase the round trip time by

$$\Delta t = \int (\mathbf{\Omega}\times\mathbf{r}).d\ell/c^2 = \mathbf{\Omega}.\int \mathbf{r}\times d\ell/c^2.$$

But the projection, on a plane normal to the axis of rotation, of the area within the path is $A = \tfrac{1}{2}\Omega^{-1}\mathbf{\Omega}.\int \mathbf{r}\times d\ell$, and so

$$\Delta t = \pm 2\ \Omega A/c^2, \tag{3.4a}$$

where the upper sign applies when light traverses the path in the direction of rotation.

As an example we apply this to a ring laser in the form of an equilateral triangle of side L, with its plane at an angle θ to the axis of rotation. The area $A=0.433L^2\sin\theta$

and the unperturbed round trip time is $t=3L/c$. Thus if the laser could be arranged to oscillate simultaneously but independently in the two modes in which light propagates around the ring in opposite directions, the fractional difference between the frequencies of these modes would be

$$\Delta f/f = 2\Delta t/t = 4\Omega A/(3Lc) = 3.6276 L\sin\theta / cT \qquad (3.4b)$$

where T is the period of mechanical rotation. The two outputs applied to the same photodetector, e.g. a photo-diode, would generate a beat frequency Δf. For example with $f = 4.10^{14}$ Hz, L=1m and T=600s (10 min) the beat frequency would be about 8 kHz and readily observable. At a latitude of 52° the effect on the same laser of the earth's rotation would be a beat of only 43 Hz.

Signals sent round the equator by reflection from the ionosphere complete a circuit in about 0.134s. The Sagnac effect causes signals travelling East-about to take some 0.4 microseconds longer than signals travelling West-about. The difference, 3 parts in 10^6, is well within the resolution of atomic clocks but this effect is of little practical importance because of the vagaries of propagation through the earth's atmosphere.

3.4 Stellar Aberration

To an observer with a horizontal velocity u, rain falling with a vertical velocity v in still air, appears to fall at an angle $\tan^{-1}u/v$ to the vertical. We might similarly expect that light rays would appear to come from a different direction to an observer in motion. Because the effect was first observed as an apparent displacement of the positions of stars as a result of the earth's orbital velocity, it became known as stellar aberration.

Consider a light ray that makes an angle θ with the x axis of a frame F, and a second frame F' that moves along this axis with a velocity u. In F the x and y components of the ray velocity are $c\cos\theta$ and $c\sin\theta$. The classical result for the corresponding velocities in F' would be $c\cos\theta - u$ and $c\sin\theta$, thus in F' the angle between the ray velocity and the x' axis would be $\theta' = \tan^{-1}[c\sin\theta/(c\cos\theta - u)]$. This can be expressed as

$$\sin\theta' = \sin\theta/(1-2u\cos\theta/c + u^2/c^2)^{1/2} \quad (3.5a)$$

and, if u/c is small, this becomes

$$\sin\theta' = \sin\theta(1+u\cos\theta/c) \quad (3.5b)$$

which gives

$$d\theta = \theta'-\theta = u\sin\theta/c. \quad (3.5c)$$

A relativistic calculation would have given

$$v_x' = c[c\cos\theta - u]/[c-u\cos\theta] \quad (3.6a)$$

$$v_y' = c^2\sin\theta/[\gamma(u)(c-u\cos\theta)], \quad (3.6b)$$

where we have used $v_x{}^2 + v_y{}^2 = c^2$. The relativistic result is therefore

$$\sin\theta' = v_y'/c = \sin\theta/[\gamma(u)(1-u\cos\theta/c]. \quad (3.6c)$$

Again if u is much less than c this gives (3.5b) and (3.5c).

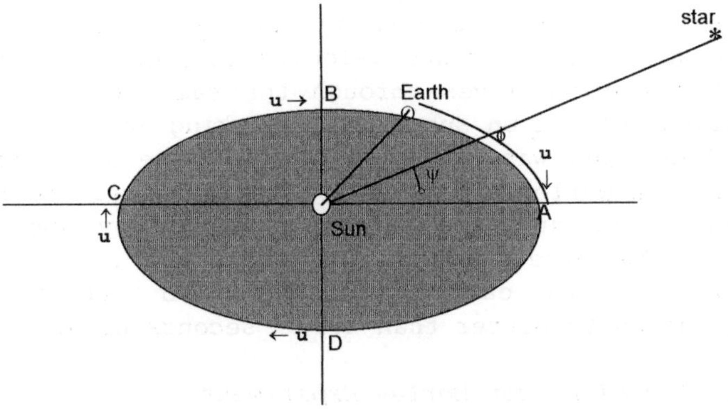

Figure 3.1 Stellar Aberration

In Figure 3.1 the angular altitude of the star above the plane of the earth's orbit is ψ and when the earth is an angular distance ϕ along its orbit, the component of its orbital velocity u parallel to the light ray is $u\cos\psi\sin\phi$. Thus the angle θ between the ray and the velocity is given by $\cos\theta = \cos\psi\sin\phi$. At A and C we have $\theta = 90^{\circ}$ and, at B and D, $\theta = \pm\psi$. The aberration at A and C is in an East-West direction and because the earth is also spinning it would be difficult to observe, but at B and D it is in a north-south direction. Between B and D the change in the star's apparent altitude is $2uc^{-1}\sin\psi$. Since $u \sim 10^{-4}c$ the change in altitude for $\psi=90^{\circ}$ is just over 40 seconds of arc.

The effect was first discovered and analysed by Bradley (1728) who was seeking a parallax effect in order to calculate the distance to a star. Since the nearest star is about 3 light years, or 3.10^{16}m, away, while the diameter of the earth's orbit is only 3.10^{11}m, the greatest angular variation in a star's altitude due to parallax would be about 10^{-5} radians or 2 seconds of arc. Furthermore the north-south parallax would be greatest at A and *C* but the aberration is greatest at B and D and, in addition, it is the same for all stars of the same altitude. Bradley realised the nature of the effect that he was observing and instead of using it to estimate the distance to the stars, he used it to derive a value for the velocity of light.

His successful interpretation, essentially in terms of a classical raindrop model, strongly suggested that starlight propagates with a fixed velocity through a fixed aether and that the earth moves through the same medium. This had a profound effect on subsequent thinking about the propagation of light and was at least part of the motivation for the Michelson-Morley experiment. The classical and relativistic results for stellar aberration differ by about $u^2/c^2=10^{-8}$ while the effect itself is about 10^{-4}. To distinguish between the two calculations one would need to measure the aberration to better than 4.10^{-3} seconds of arc.

3.5 The Michelson-Morley Experiment

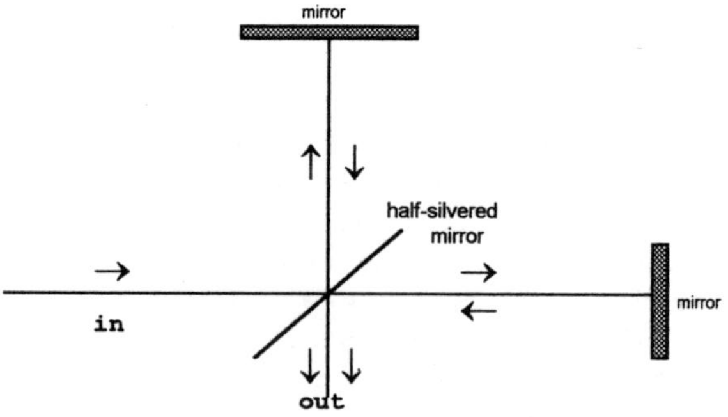

Figure 3.2 The Michelson Interferometer

The original experiment was made by Michelson alone, in 1881. Later, in a definitive and more accurate experiment, he was joined by Morley (Michelson and Morley 1887).

In the Michelson interferometer, shown in skeleton form in Figure 3.2, two orthogonal beams of light interfere. It can be oriented so that one of the arms is at right angles to the earth's orbital velocity, $u \sim 10^{-4}c$, and the other parallel to it. If the length of each arm, from the half-silvered mirror to the far mirror, is L *and* the whole instrument moves through the æther, the round trip time $t_\perp$ for light in the orthogonal arm satisfies $(\frac{1}{2}ut_\perp)^2+L^2=\frac{1}{2}(ct_\perp)^2$ which gives $t_\perp = 2L/(c^2-u^2)^{\frac{1}{2}}$. The round trip time in the parallel arm is $t_{//} = L/(c+u)+L/(c-u)$. The difference is, to second order in u/c, $t_{//}-t_\perp=(u/c)^2L/c$. At a wavelength λ the shift in the interference pattern generated by the two beams is a fraction $(u/c)^2L/\lambda$ of a fringe. With λ =500nm and arms of effective length L=20m, obtained by using multiple reflections in each arm, and with $u/c = 10^{-4}$, the expected fringe shift as the interferometer was rotated, to bring first one and then the other arm, parallel to the earth's orbital velocity was 0.4 but, to an accuracy of 1%, Michelson and Morley found **no** fringe shift.

In the years after this crucial experiment much ingenuity was expended trying to reconcile its null result with the existence of an all pervading æther through which light propagated just as sound propagates through air. Twenty years elapsed before, through the efforts of Poincaré, Lorentz and Einstein, the æther was finally abandoned.

Joos (1930) repeated the Michelson-Morley experiment and showed that, to within 1 part in 375, there was no effect due to the earth's motion . Experiments using the infra-red line in two crossed He-Ne lasers (Jaseja, et al 1964) reduced this to 1 part in 10^3, and Mossbauer experiments, reported by Isaak(1970), reduced the upper limit for the earth's velocity through the æther to 5cm/sec.

3.6 Portable Clocks

Two accurate clocks can be placed at two different places A and B at rest in the same inertial reference frame and can then be synchronised by the exchange of electromagnetic signals. It is therefore meaningful to ask whether if the clock A were to be *carried* to B it would be found to be synchronised with B on arrival. Whereas B has been at rest in an inertial frame F throughout, the clock A has been subjected to forces and accelerated to a different frame and then decelerated back to the original frame.

At some instant during its acceleration A will have a velocity v and, we shall assume, behaves like a similar *comoving* clock A^* with a constant velocity v. After an interval dt, the difference between the velocity of A and of the clock A^* will increase linearly with dt. A is now equivalent to a new comoving clock A^{**} of velocity $v + dv$. Any difference between A^* and A^{**} is proportional to $(dv/c)^2$ and so to $(dt)^2$. As dt is infinitesimal we can ignore this difference and treat A^* and A^{**} as equivalent during dt. The clock A^* records this interval dt defined in F as an interval in proper time $dt' = dt/\gamma(v)$ in its own instantaneous rest frame. When A is taken from its original position at $t=t'=0$ and arrives at B when B reads t, A will read

$$t' = \int [1 - \{v(t)/c\}^2]^{1/2} dt. \qquad (3.7)$$

The most important step in obtaining equation (3.7) was the assumption that at each instant during its accelerated motion the clock A was equivalent to a comoving clock having a constant velocity equal to the instantaneous velocity of A. The only way that we can avoid the conclusion that t' will be less than t is to assume that effects occurring during acceleration produce a shift in the properties of A that is only annulled when A is later decelerated. This is clearly unacceptable and contrary to the whole principle of relativity, for it would imply that the properties of two bodies at rest in the same inertial frame would depend on their detailed dynamic history. Fortunately we do not have to invoke such metaphysical arguments for there is direct experimental evidence that acceleration does not, of itself, produce changes other than those associated with the changed velocity. Hafele and Keating (1972) took a set of atomic clocks round the world by air, once east-about and once west-about, and compared them with standard clocks at the U.S. Naval Observatory. The results, when corrected for the Sagnac effect and also the gravitational red shift at the altitude of the flight, agreed with Eq.(3.7) to within the expected 10% experimental uncertainty. A much more precise result (Bailey et al.1977) has been obtained by studying the decay of 3 GeV muons in a storage ring. Their results agree to a few parts in a thousand with the relation $dt=\gamma(v)dt'$ even though the particles were subject to an acceleration of about 10^{19}m/s² towards the centre of the ring. It is worth noting that much larger accelerations occur for electrons in atoms (10^{23}m/s²), nucleons in nuclei (10^{31}m/s²) and in high energy electron-positron collisions (10^{33}m/s²). There are no effects that suggest that acceleration alters the passage of time in a way not accounted for in terms of the instantaneous velocity. This is also true of the dilated lifetime of muons formed at the top of the atmosphere by cosmic rays, as they fall freely to ground with no acceleration (if we accept the equivalence principle) or at most with $g=9.8\text{ms}^{-2}$.

3.7 Two Misconceptions or Apparent Paradoxes

The first misunderstanding is usually stated more or less as follows as the "Twin Paradox". Twins A and B are at rest next to each other in an inertial frame and then A sets off at high speed u on a long voyage into space and back. As far as A is concerned the voyage lasted, let us say, 30 years and once A had finished accelerating he noted that his velocity, relative to B, was 0.8c. Apart from the brief initial period of acceleration, the equally brief period of deceleration and acceleration when he turns round, and the final deceleration, he has maintained a velocity 0.8c either away, from or towards B. A finds, to his surprise, that B died and was buried 49 years after he left him.

We have by now discussed time dilation so often that we were prepared for this result and there seems to be no paradox. Paradox is usually introduced by claiming that the situations of A and B are identical; the motion of B relative to A is, apart from the sign of the velocity the same as the motion of A relative to B. Therefore we might equally argue that A should have aged more than B. No such symmetry exists. Their biological clocks were compared in the inertial frame in which B has been at rest throughout, whereas A has accelerated from one inertial frame to another and another and back, and spent appreciable time in these different frames. Although the result is strange it is not a paradox, there is no inconsistency, merely a conflict with our intuitive ideas derived from experience of much slower motion. (See problem 3.16 for an alternative approach.)

The second case, the Rod Paradox, is less familiar. A rod, whose length in its rest frame is L, moves along the axis of a substantial tube of length $0.9L$ at rest in the laboratory. The velocity of the rod relative to the laboratory frame is $0.6c$ and so its length, observed at a single instant in the laboratory, is Lorentz contracted to $0.8L$, which is less than the length of the tube. The event represented by the entry of the trailing end of the rod into the tube occurs **in the laboratory frame** before the event represented by the emergence of the leading end from the tube. In the laboratory frame there is a period when the rod is entirely within the tube and this period could be used to plug both ends of the tube with substantial stoppers, thus trapping a rod of length L in a tube of length 0.9L without either end of the rod touching the stoppers. Eventually the leading end of the rod will reach the front stopper and after this the tube can be opened with the rod inside. The result seems even stranger if we remember that in the rod's rest frame the tube is only $0.8 \times 0.9L = 0.72L$ long.

We reconcile ourselves to this result by noting that in the rod's rest frame the two plugs were not inserted simultaneously. The plug at the rear end was inserted later, and the delay is $0.9\gamma(u)uL/c^2$. During this time the tube moves a distance $0.9\gamma(u)u^2L/c^2=0.405L$ relative to the rod, and this is greater than the distance $(1-0.9/\gamma(u))L = 0.28L$, by which the tube falls short of the length of the rod. Indeed the spare length is $0.9\gamma L-L = 0.125L$. Thus, provided that it keeps moving, the rear end of the rod will have entered the tube before the rear plug is inserted (as seen in the rod's rest frame). However (again in the rod's rest frame) **before** this happens, the front end of the rod will have struck the front plug and been brought to rest relative to the tube. A shock wave then propagates back through the rod eventually bringing it all to rest relative to the tube, but this wave cannot reach the rear end of the rod in a time less than L/c, and by then the rear plug, which was inserted no later than $0.9u\gamma(u)L/c^2 = 0.675L/c$ after the front of the rod hit the front plug, is in place. Thus we end up with a highly strained rod trapped within the closed tube. This example leads to no paradox unless we assume that, because in the laboratory frame the **moving** rod is completely inside the tube, it is also inside the tube in the sense in which we might say that "the car is in the garage". We do not usually refer to the car as being in the garage when it is crumpling against the far wall at 60 let alone 400 million miles per hour. Discussing these two misunderstandings is not entirely a waste of time if it teaches us to beware of importing our everyday intuitions about space and time into discussions of motion at velocities approaching c.

3.8 Acceleration and Time Keeping

It is clear that we will eventually have to consider systems, or bodies, whose velocity changes, even if their initial and final velocities are constant. Thus we cannot completely avoid discussing acceleration, though we can often get useful results by considering only the initial and final states, usually in terms of laws such as energy conservation. We are familiar with this procedure in thermodynamics, where a great deal of useful practical information can be obtained simply by considering relations between equilibrium states, without trying to discover how a system gets from one equilibrium state to another.

We can also to some extent evade these questions (as we did in discussing the twin paradox) by assuming that any interludes of accelerated motion are so short compared with the intervals during which all velocities are constant that their effects can be ignored, but, as we shall shortly show, this subterfuge is unnecessary. There is, in principle, no difficulty in treating acceleration within the framework of special relativity though it tends to become rather tedious. But even so it is worth doing because it is the physical basis of the theory of general relativity, in which a gravitational field is treated as equivalent to acceleration of the frame of reference.

As the simplest possible example, consider the rates of clocks at different points in a constantly accelerated frame of reference. For convenience we will consider motion in only one dimension. The basic principle underlying our discussion is that *infinitesimal* time intervals are recorded by an accelerated clock in the same way as by a clock in a *comoving inertial frame* that has a constant velocity, the same as the *instantaneous* velocity of the accelerated clock. At any instant an accelerated clock is recording proper time in its own instantaneous rest frame. If two events occur at the position of an accelerated clock and are separated by dt' as recorded by this clock, then in an inertial frame in which this clock has instantaneously a velocity $v(t)$, their space and time separations are

$$dx = v\gamma(v)\,dt' = v\,dt \quad \text{and} \quad dt = \gamma(v)\,dt'. \tag{3.8}$$

Notice that $(c\,dt)^2 - (dx)^2 = (c\,dt')^2$ so that the proper time dt' corresponds to the usual invariant interval. There is, in fact, a problem hidden within this glib statement. Can we really be sure that the forces associated with acceleration do not alter the behaviour of even an atomic clock? We will return to this later and consider the calculable effect (Stark shift) associated with acceleration due to an electric field (see problem 3.11.).

We must now define *uniform acceleration*. If a particle is released from an accelerated body it remains at rest in the inertial frame that was comoving at the instant of release. The acceleration of the body relative to this inertial frame can be determined by measurements of the separation between the body and the released particle. Let the comoving frame, at time T registered in the accelerated frame, (i.e. at the proper time at the particular clock in the accelerated frame) be F^T. The proper time interval in this comoving <u>inertial</u> frame is dT and so the acceleration in F^T is

$$a = d^2x_T/dT^2 \tag{3.9}$$

and this is the acceleration experienced by the body in terms of inertial forces; or better, it is its acceleration with respect to the current comoving inertial frame. We will assume that the acceleration a defined by (3.9) is constant. If, at this instant T of proper time, the body's velocity relative to a constant inertial frame F is v, then dx_T can be transformed from the comoving frame, to dx in F and, from the relation $dx = \gamma(v)[dx_T + v\,dT]$, we obtain

$$d^2x/dT^2 = \gamma(v)\,d^2x_T/dT^2 = a\gamma(v) \tag{3.10a}$$

and
$$dx/dT = v\,\gamma(v) \tag{3.10b}$$

where
$$v = dx/dt, \tag{3.10c}$$

and t is the time in the constant inertial frame F. We also have

$$dt/dT = \gamma(v), \tag{3.10d}$$

and these equations together determine the motion of the body relative to F. If we choose F so that at $t=T=0$ the body is at rest at $x=0$ then it is easy to verify that

$$x = c^2[\cosh(aT/c)-1]/a, \quad t = (c/a)\sinh(aT/c),$$
$$v = c\tanh(aT/c) \tag{3.10e}$$

satisfy these equations. We can eliminate T to obtain

$$x = (c^2/a)[(1 + a^2t^2/c^2)^{1/2} - 1]. \tag{3.10f}$$

As long as $at<<c$ this leads to the classical results $x=\frac{1}{2}at^2$ and $v=at$, but as at increases we get $x\to ct$ and $v\to c$. Motion which is experienced as a constant acceleration *by a body*, does not appear as uniformly accelerated motion to an observer in a constant inertial frame. In this frame the acceleration, initially constant, gradually diminishes as v approaches c.

We have so far discussed the behaviour of the accelerated body in terms of a single clock and ignored any spatial extension of the body. At each instant T registered by this clock (which we take to be at the origin in the body) there is a definite comoving inertial frame, and these successive comoving frames are each in the same spatial relation to the extended body, so that the spatial coordinate systems in the comoving frames can be regarded as the instantaneous coordinate systems associated with the accelerating body. At T, a point in the body can be specified by its coordinate x_T in the current comoving frame F^T. At this point, specified by x_T in the comoving frame, the time of an event can be specified in terms of the clock at the origin in the body and in particular we can say that this event occurs at T. The origin of the body at this instant is at (x,t) in the constant inertial frame F and the event at x_T will be displaced from (x,t) by $\Delta x = \gamma(v)x_T$, and by $\Delta t = v\gamma(v)x_T/c^2$. Since $v = c\tanh(aT/c)$, we have $\gamma(v) = \cosh(aT/c)$, and thus $\Delta x = x_T\cosh(aT/c)$ and $\Delta t = x_Tc^{-1}\sinh(aT/c)$. The event that occurs at T and x_T in the body frame, therefore occurs at

$x=(c^2/a+x_T)\cosh(aT/c)-c^2/a$ and at $t=(c/a+x_T/c)\sinh(aT/c)$, in the constant inertial frame F. A neighbouring event at $T+dT$ and x_T+dx_T in the body frame occurs at $t+dt$ and $x+dx$ with

$$dx = dx_T\cosh(aT/c) + dT(c+ax_T/c)\sinh(aT/c),$$
$$dt = dx_T c^{-1}\sinh(aT/c)+dT(1+ax_T/c^2)\cosh(aT/c),$$

and so the invariant interval, which, in the constant inertial frame F, is $(d\sigma)^2=c^2(dt)^2-(dx)^2$ becomes

$$(d\sigma)^2 = c^2(dT)^2[1 + ax_T/c^2]^2 - (dx_T)^2, \qquad (3.11)$$

in the accelerated frame.

If T_x is the time registered by a clock at x_T, the invariant interval is $c^2(dT_x)^2-(dx_T)^2$ thus, if we set $dx_T=0$ in equation (3.11) we see that the interval of proper time at x_T, i.e. the time read by a clock at x_T, is related to the interval of proper time dT read by a clock at the origin by

$$dT_x = dT(1 + ax_T/c^2), \qquad (3.12)$$

We therefore conclude that clocks at different places in the *accelerated non-inertial* body frame are *not equivalent*, even though their design may be identical. A clock at the origin only measures proper time intervals between events occurring at the origin. It does not measure proper time at other points in the accelerating frame. This is of course quite different from the situation in an inertial frame where there is a common proper time throughout the frame. The situation in the accelerated frame is shown in Figure 3.3.

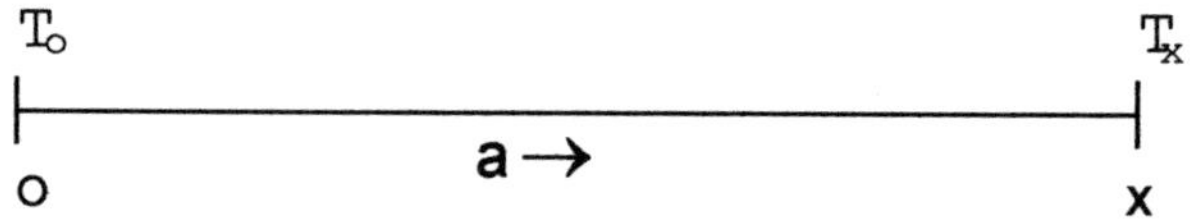

Figure 3.3 Time and Space in an Accelerated Frame

When the clock at the origin 0 records an interval dT_o between two events occurring at x, the clock at x records dT_x which from equation (3.12) is greater than dT_o. The clock at 0 runs slow in terms of proper time at x.

We may gain a further insight into the meaning of this result by considering an atomic clock at the origin 0 emitting light of frequency f_o that is received at x and compared with a similar clock at x. The light takes a time, which is about x/c, to reach x and by this time the velocity of the body and therefore of the clock at x has increased by ax/t and so the light received at x will have suffered a Doppler shift. To first order in ax/c^2 its frequency is $f_0(1-ax/c^2)$. Thus the clock at O appears, at x, to be running slow compared to the local clock. Equally, light emitted at x and received at O will have been Doppler shifted to a higher frequency and seen from O the clock at x appears to be fast. Both results agree with Eq.(3.12).

There is no simple way in which clocks at different points in a non-inertial frame can be described as synchronised or even running at the same rate. Nevertheless a standard clock at any one point reads proper time at that point, and as long as all the clocks are of identical design, it will always be possible to relate readings at different places, although in any but the simplest cases the calculation will be tedious in the extreme.

3.9 The Gravitational Red Shift

The rates of clocks in accelerated frames of reference are of no great practical interest. The importance of the calculations described in the last section arises from the equivalence principle of general relativity. This asserts that a **uniform** gravitational field present throughout a system is equivalent in all its effects to a uniform acceleration of the frame of reference. Thus the downwards gravitational force in the laboratory means that a frame of reference fixed in the laboratory must be regarded as having an upward acceleration a=9.81 ms^{-1}. This idea dates back to Boltzmann(1904). Though Boltzmann was thinking entirely in terms of classical mechanics his textbook had a profound influence on the way that mechanics developed in the 20th century. Einstein's general relativity adds to Boltzmann's idea the notion that the effect of gravity on light is also equivalent to an acceleration of the coordinate system. It leads to further problems because energy, even energy in the electromagnetic field or the gravitational field, has mass and itself experiences and generates a gravitational field.

If, in Eq.(3.12), we replace a by the acceleration g due to gravity and remember that a is then upwards, we can see that a clock at the origin runs slow when compared with a clock at a height x_T above it. Light emitted from deeper down in a gravitational field appears to be shifted in frequency towards the red end of the spectrum. A clock at sea level and one at the top of Mt. Everest differ by only about one part in 10^{12} and, although this is within the resolution of atomic clocks, errors due to the red shift are not normally a problem in a laboratory only 10m high. The gravitational shift has been frequently invoked to explain observed red shifts in the spectra of light from stars but the first quantitative experiments were not made until 1965 (Pound and Snider 1965), using the Mossbauer effect (see #7.7) and a 14.4 keV γ ray from ^{57}Fe, to measure the red shift to 1% accuracy with a height difference of only 22m.

3.10 The Thomas Precession

In discussing the internal motion of a system such as an atom, an ion, a molecule, a nucleus or indeed any entity with internal structure it is usually most convenient to work in terms of the rest frame of the centre of mass, even if the part of the system under discussion is being accelerated relative to the laboratory frame in which the system is actually being observed or described, so that one has to deal with a succession of comoving inertial frames.

We saw in section (2.6) that if a frame F' has a velocity $\mathbf{u}$ relative to a frame F and another frame F" has a velocity $\mathbf{u}+d\mathbf{v}$ relative to F, and the origins and axes of F and F' coincide at $t=t'=t''=0$, as do the axes of F" and F', then the axes of F" are rotated relative to the axes of F by

$$d\theta = [\gamma^2(u)\,d\mathbf{v}\times\mathbf{u}]/[c^2(\gamma(u)+1)] \ . \tag{3.13}$$

If the frames F' and F" are the two successive comoving inertial frames which coincide with the rest frame of the part of the system under discussion at t and $t+dt$ (relative to F) and $\mathbf{a}$ is the acceleration of the system relative to F, we have $d\mathbf{v}=\mathbf{a}dt$ and the axes rotate in the rest frame at an angular velocity

$$\mathbf{\Omega}_T = d\theta/dt = \gamma^2(u)\,\mathbf{a}\times\mathbf{u}/[c^2(\gamma(u)+1)]. \tag{3.14}$$

As a result any vector property $\mathbf{s}$ of the moving system that remains fixed in <u>its own</u> rest frame acquires a time rate of change

$$d\mathbf{s}/dt = \mathbf{\Omega}_T \times \mathbf{s}, \tag{3.15}$$

in addition to any other change in **s** that may be due to forces acting within the system. If the system itself is rotating with a constant angular velocity $\boldsymbol{\Omega}_o$ so that $\mathbf{a}\times\mathbf{u}$ and therefore $\boldsymbol{\Omega}_T$ in Eq.(3.14) are constant and parallel to $\boldsymbol{\Omega}_o$, then equation (3.15) describes a precession of **s** about the axis of rotation with an angular velocity $\boldsymbol{\Omega}_T$.

The effect is known as the Thomas Precession. It is of most interest as a precession of the electron spin in atoms, where it combines with the precession due to electromagnetic forces acting on the electron's magnetic moment. Note that the Thomas precession is not a dynamic effect due to forces, but rather an essentially kinematic effect which is determined by the space-time relations implicit in the Lorentz transformation. We will return to the significance of the Thomas precession in atomic physics, and to its role in a precise experimental test of relativity, in Chapter 9.

With u as the earth's orbital velocity and Ω_O as its angular velocity about the sun, the Thomas precession of the axis of the earth's spin is less than once every 200 million years. There may, of course, be other influences that can cause a faster precession.

3.11 Discussion

The effects considered in this chapter have been entirely concerned with the description of space, time and motion in terms of the Lorentz transformation, and although some effects such as time dilation and its practical verification through the enhanced lifetimes of fast muons, are strange and outside our normal experience of temporal relations, we might reasonably feel that if this were all there were to relativity, its present status in theoretical physics was grossly over-rated. Nevertheless this status is justified. Its dynamical consequences; not just the celebrated formula $E=mc^2$, but all its other dynamical effects, are continually felt in particle physics, nuclear and atomic physics and indirectly in chemistry and biology. These consequences also provide a sustained, and in many cases extremely precise, validation of special relativity, much as the communication industry provides a continual daily demonstration of the precision of Maxwell's equations.

Problems

We begin with three problems which, since the velocities are small compared with *c*, involve only the classical Doppler effect.

3.1 Two fighter planes approach each other with a relative velocity of 2000km/hour. One pilot observes the other, using a Doppler radar operating at 50GHz. What frequency shift does he observe?

3.2 The proximity fuses used in anti-aircraft missiles rely on a radio signal reflected from the target which beats with the outgoing signal. Why is the fuse set to fire when the beat frequency is zero?

3.3 The diaphragm of a loud speaker vibrates with an amplitude of 2mm at 1000Hz. Given that the velocity of sound in air is about 300m/sec. would you expect any significant acoustic distortion due to the Doppler effect?

3.4 A static electric field varying with x as $\sin(2\pi x/\ell)$ is set up using fixed electrodes, and an ion of velocity u parallel to x traverses the field. What is the frequency of the field in the ion's rest frame? If the smallest practical value of ℓ is 1mm, what ion velocity would be needed to induce an atomic transition at 500GHz between two nearby levels?

3.5 The wavelength of a spectral line from a distant star is found to be twice the wavelength of the same line in the laboratory. What is the velocity of recession of the star?

3.6 Two stars have equal intrinsic luminosity but one recedes at a velocity $c/3$ and the other at $c/2$. What is the ratio of the frequencies of the received photons and what is the ratio of the received intensities?

3.7 A particle travels 270mm in 1 ns. What is the elapsed proper time for this flight?

3.8 Using the simple Bohr model find the frequency of the Thomas precession in the ground state of the hydrogen atom.

3.9 Spectral lines from a distant galaxy can be identified as part of the spectrum of helium but their wavelengths are all 10% longer than the corresponding lines in the laboratory spectrum. What is the velocity of recession of the galaxy?

3.10 Pulses of electrons of velocity 0.8c pass through an aperture at intervals of 10^{-9} s. How far apart are they in their rest frame?

3.11 An electric field produces different effects on different atoms and on different spectral lines from the same atom. How do these observations affect attempts to explain the behaviour of portable clocks in terms of forces acting on them when they are set in motion or brought to rest?

In an electric field E (V/m) the Stark effect produces fractional shifts in spectral frequencies of about $10^{-20} E^2$. Estimate the Stark effect on the lines of singly ionised magnesium of a field sufficient to give the ion an acceleration of 1000g. (mass of the Mg atom, 4.10^{-26} kg)

3.12 The angular diameter of the sun is 32 minutes of arc, the stellar aberration for stars vertically above the plane of the earth's orbit is 40 seconds of arc and c=3.10^8 m/s. Estimate the gravitational red shift that would be observed, in an earth based measurement, of a spectral line emitted by an atom at the sun's surface.

3.13 The shape of a line in a spectrum can be described by the intensity $J(\lambda)d\lambda$ in an interval $d\lambda$ at a wavelength λ. Show that if atoms of mass M in a vapour at a temperature T emit a narrow line at λ_o, the Doppler broadened line shape will be $J(\lambda) = J_o \exp[-Mc^2(\lambda-\lambda_o)^2/2kT\lambda_o^2]$ where J_o is a constant. The sodium (atomic mass 23) D lines occur at 589.0 and 589.6 nm. Compare the width due to Doppler broadening in a discharge at 500K with the separation of the two lines.

3.14 What is the frequency of a 14.4 keV γ-ray and what was the expected frequency shift in Pound and Snider's red shift experiment?

3.15 Electrostatic forces cause a beam of electrons to diverge. Explain why the divergence becomes negligible for very energetic electrons as their velocity tends to c .

3.16 A rocket, carrying an observer X, passes an observer Y with a velocity v at a time recorded as t_x by X and as t_y by Y. At a time T_x (recorded by X) X is passed by another rocket borne observer Z, travelling towards Y with a velocity $-v$. As they pass Z records T_z. Z then passes Y as he records t_z and as Y records T_y. Each observer carries an accurate copy of an atomic clock. Show that $T_y - t_y = \gamma(v)[T_x - t_x + t_z - T_z]$. Comment on the connection between this result and the twin paradox. Comment also on how it relates to the many attempts to dismiss the paradox as an effect of acceleration or gravity. Would it make any difference to your comments if the three clocks had all been made in Neasden? How does the lack of symmetry between X and Z on the one hand, and Y on the other hand, arise?

CHAPTER 4

CLASSICAL DYNAMICS

Newton's first law states that in the absence of external influences a body moves with a constant velocity along a straight line. This is a purely *kinematic* law. *Dynamics* begins when we try to describe how external influences cause deviations from Newton's first law.

4.1 Mass, Momentum and Force

We certainly have a lively subjective appreciation of the difference between being struck by a tennis ball travelling at 20 m s^{-1} and a brick travelling at the same speed. This gives us a rudimentary notion of the concept of *inertial mass*. For a quantitative definition of mass we invoke the law of *conservation of linear momentum*. This is the fundamental law of *dynamics*.

To every body we can assign a numerical property, its mass m, so that, when a group of bodies with masses m_k and velocities $\mathbf{v}_k$, subject to <u>no external influences</u> from outside the group, interact and collide, the vector sum of the linear momenta $\mathbf{p}_k = m_k\mathbf{v}_k$, is the same before and after their interaction.

This clearly assumes both Newton's first law and that we know what we mean by free, non-interacting particles. We give a brief discussion in appendix A9.

In special relativity we will retain the law of conservation of momentum but to do so we will have to define *rest mass*, by taking $\mathbf{p}_k = m_k\mathbf{v}_k$ only as the limit when $|v_k|/c$ tends to zero. For v/c finite the relation becomes, as we shall see later, $\mathbf{p} = m\mathbf{v}\gamma(v) \equiv m\mathbf{v}(1-v^2/c^2)^{-\frac{1}{2}}$.

Force $\mathbf{F}$ can be defined next, in terms of the rate of change of momentum of the body on which it acts. For a single mass m of velocity $\mathbf{v}$ and momentum $\mathbf{p}$

$$\mathbf{F} = d\mathbf{p}/dt = d(m\mathbf{v})/dt, \qquad (4.1a)$$

and, in classical mechanics, where the mass m is constant, this becomes

$$\mathbf{F} = m d\mathbf{v}/dt. \qquad (4.1b)$$

4.2 Work and Energy

Though it is tedious to exert a force for any length of time it is not especially tiring, unless the object on which the force is being exerted moves in response to the force. This introduces us to the idea that moving the point of application of a force in the direction of the force involves effort or **work**. This idea is made quantitative and definite by considering the work dW done when the point of application of a force $\mathbf{F}$ is moved through $d\mathbf{r}$,

$$dW = \mathbf{F}.d\mathbf{r}. \qquad (4.2a)$$

The scalar product of the two vectors $\mathbf{F}$ and $d\mathbf{r}$ can be expressed in terms of their Cartesian components, using the double suffix summation convention, as

$$dW = F_k dr_k. \qquad (4.2b)$$

At the same time we have

$$dp_k = F_k dt, \qquad (4.3)$$

and these two equations yield the important relation

$$v_k \equiv dr_k/dt = \partial W/\partial p_k. \qquad (4.4)$$

Because this does not involve the relation between $\boldsymbol{p}$ and $\boldsymbol{v}$ it is valid in both classical and relativistic dynamics and, for that matter, it is also valid in quantum mechanics; so that the classical velocity $dE/dp = dhf/dhk = df/dk$ of a particle is the group velocity of its associated de Broglie wave, of frequency f and wave number k.

When work is done on a system its **energy** is increased, and, when the effect of a force is to increase the momentum of a body of mass m, we have $dp = Fdt$, $vdp=Fvdt=Fdr=dW$, and so the change in the **kinetic energy** T of the particle is

$$dT = vdp = mvdv = d(\tfrac{1}{2}mv^2) \qquad (4.5)$$

giving the familiar expression $\frac{1}{2}mv^2$ for the kinetic energy of a particle. Note that we can also express the kinetic energy as $p^2/2m$, and that $v_k = \partial T/\partial p_k$.

4.3 Newtonian Forces

This is about as far as we can get without making some sort of assumption about the nature of the forces acting between two or more bodies. In Newton's time, and for some centuries thereafter, the focus of attention in theoretical dynamics was on planetary motion and gravitational forces.

These forces obey an inverse square law: they act along the line of centres between the two bodies and action equals reaction. Thus the force between two bodies of masses *m* and *M* at **r** and **R** can be expressed as

$$\mathbf{F}_{mM} = -\mathbf{F}_{Mm} = -GmM(\mathbf{r} - \mathbf{R})/|\mathbf{r}-\mathbf{R}|^3 \qquad (4.6a)$$

where $\mathbf{F}_{mM}$ is the force acting on m due to the presence of M, $\mathbf{F}_{Mm}$ is the force acting on M due to m and *G* is a universal constant, 6.672×10^{-11} $\mathrm{Nm^2kg^{-2}}$. This force is conservative and so can be expressed as the negative gradient, $-\mathbf{grad}\Phi$ or $-\nabla\Phi$, of a potential energy function $\Phi(\mathbf{r})$. We then have

$$\mathbf{F}_{mM} = -\mathbf{grad}_r[-GmM/|\mathbf{r} - \mathbf{R}|], \qquad (4.6b)$$

where the subscript r on **grad** indicates differentiation with respect to **r**, and not **R**.

The peculiarly simple features of this force: it is central, conservative and leads to equal action and reaction at all instants, profoundly influenced the development of dynamics and caused notions derived from its special properties to become embedded in the fundamental theoretical structure of dynamics; in particular it led to the laws of *conservation of energy* and angular momentum. Thus, even though electromagnetic forces and the forces acting in collisions between macroscopic bodies are not Newtonian, we find it useful to employ concepts associated with Newtonian forces in discussing these interactions. For example, although angular momentum and energy are patently not conserved in glancing collisions between snooker balls, we nevertheless retain these two conservation laws and attribute the ensuing discrepancy between experiment and theory to the storage of angular momentum (spin) and energy (heat) within the balls themselves. We will discuss why this is useful later, here we fix our attention on the dynamics of a set of structureless particles interacting only through Newtonian forces. We also make the further assumption that the forces are additive. The force acting on the jth particle is the sum over k of the forces $\boldsymbol{F}_{jk}$ due to each of the other particles acting separately.

We consider first the conservation of linear momentum when the forces are Newtonian. To yield a definition of mass this conservation law had only to apply between the initial and final states, when the bodies were separated and not interacting. Thus we have so far only used it in a form which may be written as $\Delta\mathbf{P} = \int(\sum d\mathbf{p}_n/dt)\,dt = \int\sum \mathbf{F}_n dt = 0$. But if the forces are additive we have $\sum \mathbf{F}_n = \Sigma_m{}'\Sigma_n \mathbf{F}_{nm}$ where $m \neq n$, and we can express this as $\frac{1}{2}\Sigma_m{}'\Sigma_n(\mathbf{F}_{nm}+\mathbf{F}_{mn})$. Because action equals reaction, this is a sum of zeros. Thus we arrive at the much more restrictive law

$$d\mathbf{P}/dt = \Sigma_n d\mathbf{p}_n/dt = 0 \tag{4.7}$$

and the total linear momentum is constant *throughout* the motion. This has the further consequence that the velocity of the centre of mass is constant, for if $M = \sum m_k$ is the total mass, the position of the centre of mass is

$$\mathbf{R} = \left\{ \Sigma_n m_n \mathbf{r}_n \right\} / M \tag{4.8a}$$

and

$$\mathbf{V} = d\mathbf{R}/dt = \{\Sigma_n d/dt(m_n\mathbf{r}_n)\}/M = \{\Sigma_n m_n\mathbf{v}_n\}/M = \mathbf{P}/M. \tag{4.8b}$$

This result, which also holds in relativity but with a rather different interpretation, means that when we are primarily concerned with events during the interaction, we can isolate the centre of mass motion, and then remove it by a change to a new inertial frame of reference.

4.4 *Conservation of Energy*

With the force written as $\mathbf{F} = -\,\mathbf{grad}\Phi$ the equation of motion of a particle is $d\mathbf{p}/dt = -\,\mathbf{grad}\Phi$ and the rate of change of the kinetic energy T is

$$dT/dt = \mathbf{v.F} = -\mathbf{v.grad}\Phi = -d\Phi/dt.$$

Thus throughout the motion the *total energy*

$$E = T + \Phi \tag{4.9}$$

remains constant. When this is applied to a group of particles the potential Φ will depend on all the particle coordinates $\mathbf{r}_k$ and the force acting on the kth particle is

$$\mathbf{F}_k = -\,\mathbf{grad}_k\Phi(\mathbf{r}_1, \mathbf{r}_2, \ldots \mathbf{r}_k \ldots) \tag{4.10a}$$

where $\mathbf{grad}_k$ indicates that the differentiation is with respect to $\mathbf{r}_k$ alone. We also have

$$d\Phi/dt = \mathbf{v}_1.\mathbf{grad}_1\Phi + \mathbf{v}_2.\mathbf{grad}_2\Phi + \text{etc.} \tag{4.10b}$$

thus

$$-d\Phi/dt = \Sigma_k(-\mathbf{v}_k.\mathbf{grad}_k\Phi) = \Sigma_k\,\mathbf{v}_k.\mathbf{F}_k = \Sigma_k dT_k/dt \tag{4.10c}$$

where T_k is the kinetic energy of the kth particle.

In the initial and the final states, with the particles well separated, the potential energy term will be constant and can be taken to be zero. Thus

$$E_{\text{final}} = T_{\text{final}} = E = T_{\text{initial}} = E_{\text{initial}} \,. \qquad (4.10d)$$

In classical mechanics the total mass is implicitly assumed to be conserved. Even in a collision between two macroscopic bodies in which one of the bodies breaks up or adheres to the other the total mass remains unchanged. This conservation law is retained in relativity although it acquires a very different meaning.

Because M and the velocity $\mathbf{V}$ of the centre of mass are both constant the term $T_o = \frac{1}{2}MV^2$ is also constant. The remaining energy, the energy in the centre of mass system,

$$E_{\text{cm}} = E - T_o = T_{\text{initial}} - T_o \qquad (4.11)$$

must therefore be a further constant. Notice that its definition as $T_{\text{initial}} - T_o$ is independent of the conservative nature of the forces and depends only on the constancy of $\mathbf{V}$, (the velocity of the centre of mass) which, in turn, depends only on action equalling reaction, and the constancy of the total mass. If the forces are not conservative in the interaction region, the energy available to be dissipated or converted to internal energy or cause chemical or nuclear reactions, is the centre of mass energy E_{cm}.

4.5 Angular Momentum

The angular momentum of a particle of linear momentum $\mathbf{p}$ at a vector position $\mathbf{r}$ relative to the origin is $\mathbf{r}\times\mathbf{p}$ about an axis through the origin, and the total angular momentum of a system of particles is

$$\mathbf{L} = \Sigma_k\, \mathbf{r}_k\times \mathbf{p}_k. \qquad (4.12a)$$

Since $d\mathbf{r}/dt$ is parallel to $\mathbf{p}$ we have $(d\mathbf{r}_k/dt)\times\mathbf{p}_k = 0$ and so

$$d\mathbf{L}/dt = \Sigma_k \mathbf{r}_k\times d\mathbf{p}_k/dt = \Sigma_k\, \mathbf{r}_k\times \mathbf{F}_k = \Sigma_k\Sigma_l\, \mathbf{r}_k\times \mathbf{F}_{kl}.$$

The contribution from a pair of particles is $\mathbf{r}_k\times\mathbf{F}_{kl} + \mathbf{r}_l\times\mathbf{F}_{lk}$ and, since action equals reaction, this is $(\mathbf{r}_k-\mathbf{r}_l)\times \mathbf{F}_{kl}$. If the forces are also central the force $\mathbf{F}_{kl}$ is parallel to $\mathbf{r}_k - \mathbf{r}_l$, and so this vector product is zero. All the terms in $d\mathbf{L}/dt$ then cancel in pairs and as a result the total angular momentum $\mathbf{L}$ is conserved:

$$d\mathbf{L}/dt = 0. \qquad (4.12b)$$

A change of origin through a displacement $\mathbf{R}$ changes $\mathbf{L}$ to

$$\mathbf{L}^* = \Sigma_k (\mathbf{r}_k - \mathbf{R}) \times \mathbf{p}_k = \mathbf{L} - \mathbf{R} \times \mathbf{P}. \quad (4.12c)$$

If **R** is a fixed vector this is also constant since **L** and **P** are constant. If, instead of being a constant vector, **R** is the position of the centre of mass then, since both **L** and **P** are constant,

$$d\mathbf{L}^*/dt = \mathbf{P} \times d\mathbf{R}/dt = M^{-1}\, \mathbf{P} \times \mathbf{P} = 0. \quad (4.12d)$$

Thus the angular momentum about the moving centre of mass is constant. If this is denoted by $\mathbf{L}_{cm}$ then, relative to an arbitrary origin,

$$\mathbf{L} = \mathbf{L}_{cm} + \mathbf{R} \times \mathbf{P}, \quad (4.12e)$$

and, since both terms on the right are constant, we can use the initial position $\mathbf{R}_o$ of the centre of mass for **R**. In this way the total angular momentum relative to a fixed origin has been separated into an internal part $\mathbf{L}_{cm}$ which, for a composite body, might be described as "spin", and an external "orbital" part $\mathbf{R}_o \times \mathbf{P}$.

4.6 Summary of Newtonian Results

If action equals reaction the total linear momentum and the velocity of the centre of mass are constant. If in addition the forces are central, both the internal "spin" and external (orbital) angular momentum are constant. If the forces are conservative and so can be expressed as the gradient of a potential Φ the energy $E = T + \Phi$ is conserved.

It will be convenient to assemble, for future reference, the conservation laws for **structureless** particles interacting through Newtonian forces. They are

$$dM/dt = d/dt\Sigma_k\, m_k = 0, \quad (4.13a)$$

$$d\mathbf{P}/dt = d/dt\Sigma_k m_k d\mathbf{r}_k/dt = 0, \quad (4.13b)$$

$$dE/dt = d/dt\Sigma_k \tfrac{1}{2} m_k (dr_k/dt)^2 + \Phi = 0, \quad (4.13c)$$

$$d\mathbf{L}/dt = d/dt\Sigma_k \mathbf{r}_k \times \mathbf{p}_k = 0 \quad (4.13d)$$

$$M\, d\mathbf{R}/dt - \mathbf{P} = 0, \quad (4.13e)$$

and this last relation can be expressed, using (4.13b), as

$$d/dt\{\Sigma_k m_k \mathbf{r}_k - t\, \Sigma_k m_k d\mathbf{r}_k/dt\} = 0. \quad (4.13f)$$

We have, so far regarded potential energy as something associated with particles, speaking of the potential energy of a particle in, for example, the earth's gravitational field, or the potential energy of a group of interacting particles. We could however equally well have regarded the potential energy as belonging to the field. This is not particularly useful in connection with a gravitational field but much more useful for another Newtonian field, the electrostatic field **E**, where the energy density of the field can be expressed as $\frac{1}{2}\varepsilon_0 E^2$. Although a Newtonian force field can only store energy, more complicated fields such as the complete electromagnetic field can store and transport momentum and angular momentum as well as energy.

In classical mechanics there is an important difference between the conservation laws for energy and angular momentum on the one hand and those for mass and momentum on the other. Only for structureless particles can we be sure that they do not store internal energy and angular momentum, so that neither energy nor angular momentum need be conserved in the external motion of composite bodies. In contrast to this the external motion will always conserve mass and linear momentum.

4.7 Composite Bodies

A body, such as a billiard ball, can be regarded as an assembly of atomic, or sub-atomic particles, interacting mainly through the Newtonian electrostatic force. We divide the assembly into two groups a and b and then consider the motions of the two centres of mass at $\mathbf{R}_a$ and $\mathbf{R}_b$.

Clearly $dM/dt = d/dt(M_a+M_b)=0$, $d\mathbf{P}/dt = d/dt(\mathbf{P}_a+\mathbf{P}_b)=0$, and $d/dt[M_a\mathbf{R}_a+M_b\mathbf{R}_b-t(M_a d\mathbf{R}_a/dt+M_b d\mathbf{R}_b/dt)]=0$, and so the three conservation laws dealing with external properties, will retain their familiar form, and changes will be confined to the remaining two laws dealing with energy and angular momentum.

We divide the kinetic energy into two parts $T=T_a+T_b$ but divide the potential energy into three parts $\Phi=\Phi_a+\Phi_b+\Phi_{ab}$, corresponding to interactions between particles within each group separately and interactions between particles in group a and particles in group b. The total energy conservation law becomes $dE/dt = d/dt(T+\Phi) = d/dt(T_a+\Phi_a+T_b+\Phi_b+\Phi_{ab}) = 0$.

Now let the position of the jth particle in group a be $\mathbf{R}_a+\mathbf{r}_{aj}$, where $\mathbf{R}_a$ is the centre of mass of group a, then the kinetic energy of group a is $T_a = \sum_j \frac{1}{2}m_{aj}[\dot{\mathbf{R}}_a{}^2+2\dot{\mathbf{R}}_a.\dot{\mathbf{r}}_{aj}+\dot{\mathbf{r}}_{aj}{}^2]$ and since $\sum_j m_{aj} = M_a$ and $\sum_j m_{aj}\dot{\mathbf{r}}_{aj} = 0$ this yields

$$T_a = \tfrac{1}{2}M_a\dot{\mathbf{R}}_a{}^2 + \sum_j \tfrac{1}{2}m_{aj}\dot{\mathbf{r}}_{aj}{}^2,$$

where dots denote d/d*t*. The first term is the kinetic energy of the external motion of group a̲ as a whole and the second term is the internal kinetic energy. This term, together with the potential energy Φ_a, yields the internal energy H_a of group a̲ and we can express the overall conservation law as

$$dE/dt = d/dt(H_a + H_b + T_{ea} + T_{eb} + \Phi_{ab}) = 0 \quad (4.14)$$

where $T_{ea} = \frac{1}{2}M_a\dot{\mathbf{R}}_a^2$ and $T_{eb} = \frac{1}{2}M_b\dot{\mathbf{R}}_b^2$ are the kinetic energies associated with the external motions of the two subsystems. Apart from the two internal energy terms H_a and H_b, the two subsystems, or extended bodies, obey the same conservation laws as two structureless particles. However the existence of these two possibly unknown terms greatly reduces the usefulness of the energy conservation law in practical dynamical calculations.

The change in the angular momentum conservation law can be obtained by a similar procedure. The overall conservation law that results is

$$d\mathbf{L}/dt = d/dt[\mathbf{S}_a + \mathbf{S}_b + \mathbf{R}_a\times \mathbf{P}_a + \mathbf{R}_b\times \mathbf{P}_b] = 0, \quad (4.15)$$

where $\mathbf{S}_a$ and $\mathbf{S}_b$ denote the internal spin angular momenta of the two extended bodies. The separation of the total conserved angular momentum $\mathbf{L}$ into an orbital part and a spin part is important in gunnery, most ball games and particle physics.

4.8 Non-Newtonian Forces

Gravity and electrostatic forces are Newtonian but magnetic forces are very distinctly non-Newtonian. Action is not equal to reaction and the force does not act along the line of centres. When the velocities of charged particles approach the velocity of light magnetic forces are comparable with electrostatic forces; and in metals, where the electrostatic effects due to conduction electrons are annulled by positive fixed ionic cores the magnetic effects of currents are usually dominant. Thus one of the principal forces operating on a macroscopic scale is non-Newtonian and nuclear forces also share this property. We must see, therefore, how the Newtonian conservation laws can be modified to accommodate non-Newtonian forces.

A clue is provided by the way that we can treat potential energy as either an attribute of the particles or as an attribute of the force field. If we are prepared to let the field store energy, there is no reason why we should not also allow it to store linear and angular momentum. In classical theory we might draw the line at ascribing mass to a field though, as we shall see, this is necessary in relativity. Our aim is to keep all the familiar conservation laws by attributing energy, linear and angular momentum and even, if necessary, mass to the field in such a way that the sums of the energies, linear momenta, angular momenta and masses of the particles and the corresponding field quantities are conserved. This is of course only useful if we can express the field quantities in terms of field variables with an independent significance, e.g. the electric and magnetic fields **E** and **B**.

As an illustration consider the situation shown in Figure 4.1, where two charged particles A and B move in perpendicular directions. The field due to A at B is zero

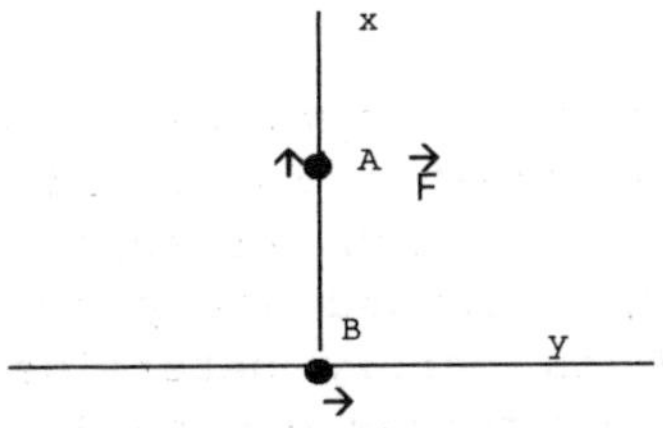

Figure 4.1 Two Moving Charges

and so there is no magnetic force acting on B, but the field due to B at A is non-zero, and since it is perpendicular to the motion of A it produces the force **F**. The non-Newtonian nature of the force is evident. However, electromagnetic theory tells us that the total forces $\mathbf{F}_A$ and $\mathbf{F}_B$ acting on the charges are related to the fields **E** and **B** by

$$\mathbf{F}_A + \mathbf{F}_B + d/dt\iiint \varepsilon_o(\mathbf{E} \times \mathbf{B})\,dV = 0.$$

We can, therefore, preserve conservation of linear momentum by ascribing a momentum density $\varepsilon_o\mathbf{E}\times\mathbf{B}$ to the field. This is, in fact, just the energy flux vector (Poynting's vector $\mathbf{E}\times\mathbf{H} = \mathbf{E}\times\mathbf{B}/\mu_o$) divided by c^2. During the interaction the field momentum is changing but initially, and finally, when the particles are far apart, the volume integral will be constant. Thus, from the initial to the final state, we have

$$\int(\mathbf{F}_A + \mathbf{F}_B)\,dt = -\,\Delta \iiint \varepsilon_o(\mathbf{E} \times \mathbf{B})\,dV,$$

where the right hand side is the difference between the initial and final values of the field momentum. In the initial and final states, unless there has been radiation, the volume integral can be separated into two parts, one associated with each particle. Let us call these two parts the field momenta $\mathbf{P}_{fA}$ and $\mathbf{P}_{fB}$, then if $\mathbf{p}_A$ and $\mathbf{p}_B$ are the particle momenta we have the simple conservation law

$$\Delta(\mathbf{p}_A + \mathbf{P}_{fA}) \quad + \quad \Delta(\mathbf{p}_B + \mathbf{P}_{fB}) \quad = \quad 0. \qquad (4.16)$$

This associates a field momentum with each charged particle as part of its total momentum. If a charged particle collides with a massive uncharged body and is brought to rest, the momentum transferred to the massive body is $\mathbf{p}+\mathbf{P}_f$. The field of a charged particle contributes to its momentum, energy and, in relativity, to its mass.

4.9 *Impressed Forces*

No physical system is completely isolated, though for many purposes we can treat a localised system as isolated, However even then it is often convenient to divide a system into parts A, B, C, etc. and study only one part, say A. The behaviour of A is partly determined by forces acting within A and partly by forces due to B, C, etc. As long as only the behaviour of A is significant, the origin of these forces is of no interest and they can be treated as *given*, or *impressed* forces. Thus, when studying proton nuclear magnetic resonance in water, we represent the entire effect of the external magnet on the specimen by the magnetic field **B** that it produces.

Impressed forces acting on a system can be treated in two ways. The external influence can be regarded as a force field acting throughout the system, or it can be regarded as producing a system of stresses over the surface of the system. We can also use a mixture of the two approaches. This would for example be appropriate in considering the flight of a golf ball from the moment that it was struck to the time that it came to rest.

Here we take the first point of view and let $\mathbf{f}(\mathbf{r})$ be the impressed force field and $\mathbf{f}_k$ be the force acting on the kth particle at $\mathbf{r}_k$. Instead of the conservation laws we have

$$\begin{aligned} d\mathbf{P}/dt &\equiv d/dt\ \Sigma_k \mathbf{p}_k = \Sigma_k \mathbf{f}_k = \mathbf{F}, \\ dE/dt &= \Sigma_k \mathbf{v}_k . \mathbf{f}_k, \\ d\mathbf{L}/dt &= \Sigma_k \mathbf{r}_k \times \mathbf{f}_k = \mathbf{C}, \\ d^2\mathbf{R}/dt^2 &= M^{-1} d\mathbf{P}/dt = \mathbf{F}/M, \end{aligned} \tag{4.17}$$

where $\mathbf{F}$ is the total impressed force and $\mathbf{C}$ is the total impressed couple, or torque, acting on the system.

It is possible for the total force to be zero while the individual forces are non-zero and in this case, though $\mathbf{P}$ and $d\mathbf{R}/dt$ will be constant, E and $\boldsymbol{L}$ may vary. If, however, the total angular momentum is separated into an internal spin component $\mathbf{S}$ and an external orbital component $\mathbf{R} \times \mathbf{P}$ so that $\boldsymbol{L} = \mathbf{S} + \mathbf{R} \times \mathbf{P}$, we see that though $d\mathbf{S}/dt$ will be non-zero we will have

$$d/dt(\mathbf{R} \times \mathbf{P}) = (d\mathbf{R}/dt) \times \mathbf{P} + \mathbf{R} \times d\mathbf{P}/dt = 0, \tag{4.18}$$

for $d\mathbf{R}/dt$ is parallel to $\mathbf{P}$ and $d\mathbf{P}/dt = 0$ if the total force is zero. Thus in the absence of external couples the external angular momentum is conserved.

This concludes our brief resumé of classical dynamics in which we have presented those concepts of particular relevance to relativistic dynamics.

Problems

4.1 The energy of a free particle cannot depend on the direction of its momentum. Show that this implies that its momentum is parallel to its velocity.

4.2 The classical equation of motion of an electron of charge $-e$, mass m and velocity $\mathbf{v}$, in an electric field $\mathbf{E}$ and a magnetic field $\mathbf{B}$, is $m d\mathbf{v}/dt = -e(\mathbf{E}+\mathbf{v}\times\mathbf{B})$. In a cylindrical magnetron there is a uniform axial magnetic field B, a co-axial cathode of radius $r=a$ and an anode of radius $r=b>a$. Electrons are emitted from the cathode with negligible initial velocities. By considering angular momentum, show that during the motion of the electrons between the cathode and anode (which is at a positive voltage V relative to the cathode) $L^* = r^2(\frac{1}{2}eB - m d\theta/dt)$ remains constant and, by considering energy conservation, that electrons only reach the anode if $V > (eB^2/8m)(b^2-a^2)^2/b^2$.

4.3 A particle of mass m with a total energy E moves in a region near the origin where the potential energy is $k|x|$ where k is a positive constant. Find the frequency of the periodic motion.

4.4 A cylinder of radius 10mm, length 50mm and mass 0.12kgm rolls without slipping down a plane inclined at 30° to the horizontal. Find the acceleration of its centre of mass.

CHAPTER 5

RELATIVISTIC DYNAMICS I

5.1 Mass and Momentum

In classical mechanics the momentum **p** of a particle of velocity **v** is m**v** where *m*, the mass, is a constant property of the particle; and the two basic laws of dynamics are the conservation of momentum and of mass. In relativity we try to retain these two laws but in a way consistent with the Lorentz coordinate transformation rather than the Galilean transformation of classical physics. The price that we pay is the replacement of the constant mass in **p** = m**v** by a mass which, though remaining characteristic of the particle, depends on its speed *v*. Since all directions in space must be equivalent, if $\mathbf{p} = m(\mathbf{v})\mathbf{v}$, then $-\mathbf{p} = -\mathbf{v}m(-\mathbf{v})$, and so *m* can only depend on the magnitude of **v**, i.e. the speed *v*. Also, because relativistic dynamics must reduce to classical dynamics at low velocities, the mass $m(v)$ must tend to a constant value m_0, the *rest mass*, as v/c tends to zero.

We have now to discover the form of $m(v)$ that preserves both the conservation laws of momentum and mass when the Lorentz transformation is used to go from coordinates in one inertial frame of reference to coordinates in another, and we do this in two stages.

Consider first two identical particles which, in the laboratory frame F, have equal, but opposite, velocities *v* and *-v* along the *x* axis. Because their speeds are equal their momenta are equal but opposite and so, whether the collision is elastic or inelastic, there will be an instant when both the particles are at rest. In a new frame F', with a velocity *u* along the line of the motion, the initial velocities are

$$v_1' = (v - u)/(1 - vu/c^2) \tag{5.1a}$$

$$v_2' = (-v - u)/(1 + vu/c^2) \tag{5.1b}$$

and because these are of unequal magnitude we cannot assume that the masses m_1' and m_2' are the same. At the instant when both particles are at rest in F their velocity is $-u$ in F'. Let M' be the inertial mass of the whole system at this instant (it is not just the sum of the particle masses at this instant, see #5.2). Momentum conservation then requires

$$m_1'v_1' + m_2'v_2' = -M'u, \tag{5.2a}$$

and mass conservation requires

$$m_1' + m_2' = M'. \tag{5.2b}$$

With M' eliminated this gives $m_1'(v_1'+u)+m_2'(v_2'+u) = 0$ and, with the aid of Eqns.(5.1a,b), this can be expressed as

$$m_1'/m_2' = (1-vu/c^2)/(1+vu/c^2). \tag{5.3}$$

This gives the mass ratio in the frame F' in terms of the velocity v in the frame F. To get all the variables in terms of F' we use the relation (see Eq.(2.21) in section 2.6)

$$\gamma(v') = \gamma(u)\gamma(v)(1-vu/c^2) \tag{5.4}$$

and thus obtain

$$m_1'/m_2' = \gamma(v_1')/\gamma(v_2'). \tag{5.5}$$

Now m_1' can only depend on v_1' and m_2' on v_2' and also, when the velocities tend to zero and $\gamma(v)=(1-v^2/c^2)^{-\frac{1}{2}}$ tends to unity, the masses must reduce to the rest mass m_0, so that we get $m_1(v_1')=m_0\gamma(v_1')$. In this special one-dimensional dynamical situation the relation between the momentum $\mathbf{p}$ of a particle (of rest mass m_0) and its velocity must be

$$\mathbf{p} = m\mathbf{v} = m_0\gamma(v)\mathbf{v}, \tag{5.6}$$

and the inertial mass is therefore

$$m = m_0\gamma(v) = m_0(1-v^2/c^2)^{-\frac{1}{2}}. \tag{5.7}$$

Consider next a single particle of rest mass m_0 and vector velocity $\mathbf{v}$ in an arbitrary direction relative to F. If we now assume that the results (5.6) and (5.7) are completely general, then in F we shall have $p_x = m_0 v_x\gamma(v)$, $p_y = m_0 v_y\gamma(v)$, $p_z = m_0 v_z\gamma(v)$, or more succinctly, $p_j = m_0 v_j\gamma(v)$ and, in a frame F' moving with velocity u along the x axis, we must, to obtain invariant equations, have $p_j' = m_0 v_j'\gamma(v')$. Since $v_1'=(v_1-u)/(1-uv_1/c^2)$, $v_2'= v_2/[(1-uv_1/c^2)\gamma(u)]$, and $v_3'= v_3/[(1-uv_1/c^2)\gamma(u)]$ and also $\gamma(v')= \gamma(v)\gamma(u)(1-uv_1/c^2)$, we see that

$$p_x'=m_0\gamma(v)\gamma(u)(v_x-u), \tag{5.8a}$$

$$p_y'=m_0\gamma(v)v_y=p_y, \tag{5.8b}$$

$$p_z'=m_0\gamma(v)v_z=p_z, \tag{5.8c}$$

and so the transverse components of $\mathbf{p}$ are unchanged.

If, in F, the total momentum $\mathbf{P}=\Sigma_n\mathbf{p}_n$ of a set of particles is conserved, then the transverse components are certainly conserved in F'. We therefore need only consider conservation of the parallel components.

Initially when all the particles are well separated

$$P' = \sum p_n' = \gamma(u)\sum m_{on}\gamma(v_n)(v_n-u) \qquad (5.9)$$

which we can express as

$$P' = \gamma(u)P - \gamma(u)Mu \ , \qquad (5.10)$$

where $M = \sum_n m_n = \sum_n m_{on}\gamma(v_n)$ is the total inertial mass in the initial state. If therefore P and M are conserved between the initial and the final state, then P' is also conserved.

Expressed in terms of the reference frame F', the total inertial mass in the initial state is

$$M' = \sum m_{on}\gamma(v_n') = \sum m_{on}\gamma(u)\gamma(v_n)(1-\mathbf{u}.\mathbf{v}_n/c^2), \qquad (5.11)$$

where, to simplify the notation we have written uv_{xn} as $\mathbf{u}.\mathbf{v}_n$. Equation (5.11) can also be written as

$$M' = \gamma(u)[M-\mathbf{u}.\sum m_{on}\mathbf{v}_n\,\gamma(v_n)/c^2] = \gamma(u)[M-\mathbf{u}.\mathbf{P}/c^2]. \quad (5.12)$$

Since $\mathbf{u}$ is a constant parameter we see that if M and $\mathbf{P}$ are conserved in F then M' and, from Eq.(5.10), $\mathbf{P}'$ are conserved in F'. Thus, from the initial to the final states, where the particles are far apart and not interacting, the two basic conservation laws are invariant under a relativistic Lorentz transformation of the coordinates and time. They are therefore consistent with the general principles of special relativity. However we must emphasise that we have only discussed the conservation laws as they apply to the initial and final states, when the particles are far apart. In effect we have only shown that the relation $\mathbf{p} = m_o\gamma(v)\mathbf{v}$ applies to free particles. We have not attempted to pursue the conservation laws into the interaction region as we did in the classical case (section 4.3). If we assume that the relation $\mathbf{p} = m_o\gamma(v)\mathbf{v}$ is valid for interacting particles we find that to preserve the conservation laws throughout the interaction region we have to attribute mass to the force fields. We postpone further discussion of this point until we have dealt with work and energy in relativity.

By considering the very special case of a head-on collision of two particles of equal speed and mass we were forced to the conclusion that we could only preserve conservation of mass and momentum if we set the inertial mass of a particle of rest mass m_o and velocity v equal to $m_o\gamma(v)$. We then assumed that this was a general relation, so that the momentum of a particle of rest mass m_o and velocity $\mathbf{v}$ would be $m_o\gamma(v)\mathbf{v}$ and showed that, as a result, if the total momentum and mass of a system of particles were conserved in one inertial frame of reference, then they would also be conserved in any other inertial frame. Eventually of course the validity of this self-consistent structure must depend on how it agrees with experiment.

5.2 Energy and Mass

We return to our original example of two equal particles moving along the same line with initial velocities v and $-v$ and then colliding and afterwards separating with velocities $-w$ and w. The total inertial mass before the collision is $M_i = 2m_o\gamma(v)$, and after the collision it is $M_f = 2m_o\gamma(w)$. If the collision is elastic $w=v$ and mass is conserved, but if it is inelastic, $w<v$ and the final mass is less than the initial mass. Apparently in inelastic collisions, we will have to abandon the conservation of mass in its simplest form. This does not spoil the arguments of the last section where we were only concerned to show that mass and momentum were conserved in F' **if** they were conserved in F.

At the moment when both masses are instantaneously at rest in F we have $m=m_o$ for both particles and so $M=2m_o$. Thus if we set $M = m_o[\gamma(v_1) + \gamma(v_2)]$ where v_1 and v_2 are the instantaneous velocities, M will not be a constant of the motion. Since we have just seen that M may not even be conserved between the initial and the final states unless the collision is elastic and energy is conserved, this suggests that conservation of mass is somehow connected with conservation of energy.

Even if the collision is elastic and finally conserves kinetic energy, the interaction between the particles is still associated with a changed potential energy during the interaction. In the interaction region the mass deficit in an elastic collision is $2m_o[\gamma(v)-\gamma(w)]$ where w (less than v) is the instantaneous speed. If v/c is small the deficit is $-\Delta M = 2m_o[\gamma(v)-\gamma(w)] \approx 2[\frac{1}{2}m_o v^2-\frac{1}{2}m_o w^2]/c^2$ and this gives $-\Delta M \approx \Delta T/c^2$ where ΔT is the decrease in kinetic energy and is therefore also the increase $\Delta\Phi$ in potential energy at this particular instant. With $\Delta M \approx \Delta\Phi/c^2$ it appears that if we want to retain conservation of mass, we will have to associate a mass Φ/c^2 with the potential or field energy Φ.

If the collision had been inelastic the energy released could either have been radiated away or gone into the internal energy of the colliding bodies, possibly as heat, or as internal strain or as vibrational or rotational energy. If, therefore, we still want to preserve the conservation of inertial mass we can only conclude that, *wherever the energy went, it took mass with it.*

5.3 Force and Work

We will retain the definition of force as rate of change of momentum so that

$$\mathbf{F} = d\mathbf{p}/dt = d/dt[m_0\mathbf{v}\gamma(v)]. \tag{5.13}$$

We will also retain the notion of work as 'force times displacement'

$$dW = \mathbf{F}.d\mathbf{r}, \tag{5.14}$$

and the further notion that the increase in the kinetic energy T of a body is

$$dT = \mathbf{F}.d\mathbf{r} = (d\mathbf{p}/dt).d\mathbf{r} = (d\mathbf{p}/dt).\mathbf{v}dt = \mathbf{v}.d\mathbf{p}. \tag{5.15}$$

In terms of components

$$dT = v_k dp_k \tag{5.16}$$

with the usual sum over k=1,2,3. This leads to the relation, common to classical and relativistic dynamics,

$$v_k = \partial T/\partial p_k. \tag{5.17}$$

Now

$$p^2 = m_0^2 v^2/(1-v^2/c^2) \tag{5.18}$$

and, solved for v^2, this gives

$$v^2 = p^2c^2/(m_o^2c^2+p^2), \tag{5.19}$$

and so, since $p_1/v_1 = p_2/v_2 = p_3/v_3 = m_0\gamma(v)$, we have

$$v_k = cp_k\,/(m_o^2c^2+p^2)^{1/2}. \tag{5.20}$$

We can now use Eq.(5.16) to calculate T in terms of p, for we have

$$T = \Sigma_k\int v_k dp_k = \int\Sigma_k cp_k dp_k/(m_o^2c^2+p^2)^{1/2}.$$

But $dp^2 = d\Sigma_k(p_k p_k)$ and so

$$T = \tfrac{1}{2}\int c\,dp^2/(m_o^2c^2+p^2)^{1/2} = c(m_o^2c^2+p^2)^{1/2} - m_0c^2. \tag{5.21}$$

If we write this as

$$T = m_0c^2[(1+p^2/m_o^2c^2)^{1/2} - 1] \tag{5.22}$$

we see that for $p \ll m_0c$

$$T \approx m_0c^2[1+p^2/2m_0^2c^2 - 1] = p^2/2m_0 \tag{5.23}$$

which is just the classical result.

By using Eq.(5.18) T can be expressed in terms of v with the result

$$T = m_o c^2[(1-v^2/c^2)^{-\frac{1}{2}} - 1] = m_o c^2[\gamma(v) - 1]$$

and this gives

$$T = mc^2 - m_o c^2 \qquad (5.24a)$$

or

$$m - m_o = T/c^2 \qquad (5.24b)$$

Thus the increased inertial mass of a moving free particle is, apart from the factor $1/c^2$, just its kinetic energy.

Since T is by definition (see Ch.4) equal to the work done by applied forces which, if they are conservative, can be derived from a potential Φ, we will have, just as in classical theory, $\Delta(T+\Phi) = 0$. If, therefore, a particle enters a region where the potential energy is $+\Phi$, its mass will decrease by Φ/c^2 and so, if we propose to retain conservation of mass, we must attribute a mass Φ/c^2 to the field.

We now see, reverting to our discussion of two colliding particles, that if we ascribe a mass Φ/c^2 to their interaction energy, the quantity $M = m_o\gamma(v_1) + m_o\gamma(v_2) + \Phi/c^2$ will be a constant of the motion. Our approximate guess in the last section is in fact exactly right. In addition, in terms of the kinetic energies T_1 and T_2 of the particles,

$$Mc^2 = m_o c^2 + m_o c^2 + T_1 + T_2 + \Phi = 2m_o c^2 + T_o \qquad (5.25)$$

where T_o is the total kinetic energy of the particles before they began to interact. Eq.(5.25) can also be written as

$$Mc^2 = 2m_o c^2 + Q \qquad (5.26)$$

where $Q = T_o = T_1 + T_2 + \Phi$ is the total (in the sense of classical mechanics) energy of the particles, and in this form the close connection between conservation of energy and conservation of mass becomes apparent.

Any composite body, from an atom to a ballistic missile, can be regarded as an assembly of simpler bodies and we see that its rest mass M can be expressed as the sum of the rest masses of its constituent particles plus their total energy in the body (divided by c^2 of course). It is tempting to suppose that each of these constituents can be analysed into smaller and smaller units, and clearly it would then become sensible to absorb all the rest mass quantities $m_o c^2$ into a new total internal energy Q and simply write

$$M = Q/c^2. \qquad (5.27)$$

This only becomes misleading if there is a point at which further subdivision is impossible, so that there is a residual rest mass part of Q not available as energy even if the whole composite body were decomposed. This objection is (at least as a general objection) unfounded for, as we know, the simplest of all sub-atomic particles (the electron and the positron) can annihilate, releasing all their rest mass energy as radiation. There is, therefore, no experimental objection to regarding Eq.(5.27) as a universal relation between mass and energy.

The inertial mass M of a body not only varies with its velocity according to $M = M_0(1-v^2/c^2)^{-\frac{1}{2}}$ but its rest mass M_0 itself can vary if the internal energy of the body changes, perhaps as a result of chemical reactions, heat or spin resulting from a collision. Any energy supplied to the body from outside must, to conserve mass, add to the mass of the body.

Although discussed earlier by both Poincaré and Einstein (1905b), the idea of a universal relation $E=Mc^2$ between mass and energy seems first to have been related to Kaufmann's (1902) experimental results (see #2.1) by G.N.Lewis (1908). Lewis also quoted Landolt as showing that any mass change accompanying a chemical reaction, with energy changes of at most a few eV, would be too small to be experimentally measurable. Einstein suggested that radio-active decay offered the best hope for experimental verification.

Nuclear reactions involve energy changes of 10^7eV, or so, and the situation is different from chemical reactions. Changes in M_0 are readily detectable, especially in light nuclei. All the available evidence from electron-positron annihilation and nuclear reactions confirms the Einstein relation $E=Mc^2$ to better than one part in a thousand.

Equation (5.26) can be written as

$$mc^2 = T + m_0c^2 = T + Q = E \tag{5.28}$$

where E is now the total energy of the body, except for its potential energy in any external field due to other bodies. The excluded potential energy term yields a mass associated with the field and thus not with the body alone but rather with its interaction with other bodies. This energy would however have to be included in E if the whole collection of interacting bodies were to be grouped together and treated as a single object. Thus the rest mass of the carbon nucleus includes the rest masses of the protons and neutrons and their energy of interaction but not the energy of their interaction with the orbital electrons; that, together with the rest masses of the electrons would be included in the rest mass of the complete atom.

The relation between $E=T+m_0 c^2$ and momentum p is, from Eq.(5.21),

$$E^2 = p^2c^2 + m_0^2c^4 \tag{5.29}$$

and this result is often regarded as the basic equation of relativistic particle dynamics. The components of the velocity can be obtained from it, using

$$v_k = \partial E/\partial p_k \tag{5.30}$$

and this leads to Eq.(5.19), $v^2=p^2c^2/(m_0^2c^2+p^2)$, and also to $\mathbf{p} = m_0\mathbf{v}/(1-v^2/c^2)^{1/2} = m_0\mathbf{v}\gamma(v)$.

When Eq.(5.29) is written as

$$E^2 - p^2c^2 = m_0^2c^4 \tag{5.31}$$

we see that, for a body of constant structure and internal energy, $E^2 - p^2c^2$ is an invariant, the same in all inertial frames of reference. The similarity to the invariant interval $c^2t^2 - r^2$ is obvious and the existence of Lorentz invariants such as this is the basis of the 4-vector treatment introduced in Chapter 8.

5.4 Summary

Before we start to use these results we recall their origin. We set out intending to retain the form of the classical laws of momentum and mass conservation, the definition of force as rate of change of momentum, the definition of work as force times displacement and the equivalence between work done on a body and a change in a function (the kinetic energy) of the variables describing the motion of the body. We then required that the dynamical laws so retained should be Lorentz invariant i.e. take the same form in any inertial frame of reference. This forced us to abandon the concept of an invariant inertial mass and to replace $\mathbf{p} = m_0\mathbf{v}$ by $\mathbf{p} = m_0\mathbf{v}/(1-v^2/c^2)^{1/2}$.

If $m = m_0/(1-v^2/c^2)^{1/2}$ is to become a general relation, then conservation of inertial mass also requires inertial mass to be ascribed to any form of energy, for if a force slows down a particle increasing its potential energy, not only does it reduce its momentum and kinetic energy but it also reduces its inertial mass.

The result of these steps, summarised by $E^2=p^2c^2+m_o^2c^4$ is a new set of dynamical equations for the motion of particles in a single frame of reference. Thus, by requiring the laws to be the same in any inertial frame, we have altered the laws that apply in a single frame and, $E^2=p^2c^2+m_o^2c^4$, for example, replaces the classical equation $T=p^2/2m$, although it reduces to this equation if p/m_oc or v/c is small.

The most remarkable feature of the new laws is the connection between mass and energy. We also note that Eq.(5.31) suggests that E must be the time-like part of an invariant quantity whose space-like part is $c\mathbf{p}$.

Problems

5.1 What is the constant quantity corresponding to L^* in question (4.2) when the velocity of the electrons becomes comparable with c?

5.2 Find the ratio of the inertial mass of an electron of velocity $0.8c$ to its rest mass.

5.3 Use the equation $E^2 = p^2c^2+m^2c^4$ to express $v = dE/dp$ and dv/dp in terms of p and m. Show that $m^* \equiv dp/dv = m\gamma^3(v)$ and express $\gamma(v)=(1-v^2/c^2)^{-1/2}$ in terms of p.

5.4 For electrons $m_oc^2 \approx 500$keV, find the velocity of an electron accelerated from rest through a potential difference of 10^6V.

5.5 What is the ratio of E to pc for electrons of total energy 10^9V?

5.6 A beam of 500keV electrons traverses a narrow gap in a cavity resonator oscillating at 10^{10} Hz with a peak voltage 10kV across the gap. How far away from the gap will electrons that crossed the gap when the accelerating voltage was at its maximum overtake electrons that crossed it when the voltage was zero?

5.7 The capacitance of a sphere of radius a is $4\pi\varepsilon_0 a$ and it was thought at one time that the electron's mass m and its rest energy m_0c^2 might be due to the electrostatic energy associated with its charge. What numerical value does this give for the radius of the electron?

5.8 How long would a 1000 Mwatt power station have to run to accelerate a mass of 1kgm to a velocity $0.6c$?

5.9 The force acting on an electron of charge e and velocity $\mathbf{v}$ in a magnetic field $\mathbf{B}$ is $e\mathbf{v}\times\mathbf{B}$. Find the radius of the orbit of an electron of energy 50 MeV in a uniform magnetic field of 1T.

CHAPTER 6

RELATIVISTIC DYNAMICS II

6.1 Radiation Pressure

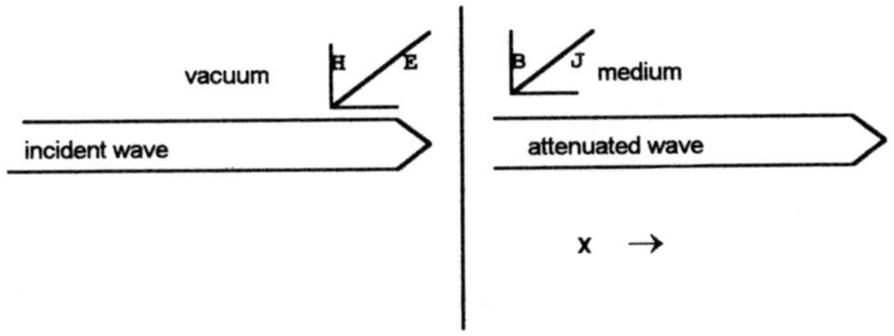

Figure 6.1 Radiation Pressure

In Figure 6.1 a plane electromagnetic wave of angular frequency ω and characterised by fields **E** and **H**, falls at normal incidence from vacuum on the plane surface of a medium, whose only difference from vacuum is a small electric conductivity $\sigma \ll \omega\varepsilon_o$, so small that we can ignore any reflection, since this would be quadratic in $\sigma/\omega\varepsilon_o$. In vacuum the wave propagates as $e^{i(\omega t-\kappa_o x)}$ where $\kappa_o=\omega/c$. In the medium κ_o is replaced (to first order in σ) by $\kappa = \kappa_o-\frac{1}{2}i\sigma Z_o$, where $Z_o=(\mu_o/\varepsilon_o)^{\frac{1}{2}}$ is the impedance of free space. Thus the amplitudes of the fields decay with distance as $e^{-\frac{1}{2}\sigma Z_o x}$. The electric field drives a current density $\mathbf{J} = \sigma\mathbf{E}$ and the electromagnetic force density will therefore be $\mathbf{f} = \mathbf{J}\times\mathbf{B} = \mu_o\mathbf{J}\times\mathbf{H} = \mu_o\sigma\mathbf{E}\times\mathbf{H} = \mu_o\sigma\mathbf{E}_o\times\mathbf{H}_o e^{-\sigma Z_o x}$. The total force per unit area acting on the medium is obtained by integrating this force per unit volume from x=0 to infinity, and this gives

$$\mathbf{F} = \mathbf{E}_o\times\mathbf{H}_o/c. \tag{6.1}$$

The rate at which momentum is brought up to unit area is thus $1/c$ times the rate $\mathbf{E}_o\times\mathbf{H}_o$ at which energy is brought up.

The relation between the energy E and the momentum p of a plane electromagnetic wave is therefore

$$E = pc. \tag{6.2}$$

This yields

$$v = \partial E/\partial p = c, \tag{6.3}$$

which need not surprise us, and the equation

$$E^2 = p^2c^2. \tag{6.4}$$

This replaces $E^2 = p^2c^2 + m_0{}^2c^4$, and corresponds to treating a photon as a particle of zero rest mass, i.e. $m_0=0$. However, if we write $p = mv = mc$, the inertial mass is still $m = E/c^2$ (as we assumed in Chapter 5).

Thus in an inelastic collision in which energy is carried off by radiation the radiation will also carry off mass.

6.2 *The Transformation of Energy*

The transformation laws for the three components of momentum are (see eqn.5.8)

$$p_x' = m_0\gamma(u)\gamma(v)[v_x-u), \quad p_y' = p_y, \quad p_z' = p_z. \tag{6.5}$$

For the energy we have $E = mc^2 = m_0c^2\gamma(v)$ so that E must transform like $\gamma(v)$ which, see Eq.(2.12), gives

$$E' = m_0c^2\gamma(u)\gamma(v)(1-\mathbf{u.v}/c^2) = E\gamma(u)(1-\mathbf{u.v}/c^2). \tag{6.6a}$$

The inverse transform obtained by changing the sign of u is

$$E = E'\gamma(u)(1+\mathbf{u.v}'/c^2) \tag{6.6b}$$

and for these to be consistent we must have

$$(1-\mathbf{u.v}/c^2)(1+\mathbf{u.v}'/c^2) = \gamma^{-2}(u). \tag{6.7}$$

We now check this: $\mathbf{u.v}' = uv_x' = u(v_x-u)/(1-\mathbf{u.v}/c^2)$, thus $(1-\mathbf{u.v}/c^2)(1+\mathbf{u.v}'/c^2)=1-uv_x/c^2+u(v_x-u)/c^2=1-u^2/c^2=\gamma^{-2}(u)$ QED.

The kinetic energy is $T = E-m_0c^2$ and so the Lorentz transformation of the kinetic energy is

$$T' = T\gamma(u)(1-\mathbf{u.v}/c^2)+m_0c^2[\gamma(u)(1-\mathbf{u.v}/c^2)-1]. \tag{6.8}$$

Obviously if T and $\mathbf{v}$ are zero in F we get $T' = m_0c^2[\gamma(u)-1]$ and this, as we might expect, corresponds to an inertial mass $m'=m_0\gamma(u)$ in F'. The classical limit of (6.8) when u/c and v/c are small is, to second order in these small quantities, $T = \frac{1}{2}m_0 v^2$, and $T' = \frac{1}{2}m_0(v^2+u^2-2\mathbf{u.v}) = \frac{1}{2}m_0|\mathbf{v}-\mathbf{u}|^2$, as we might also have guessed, since in F' the classical result for $\mathbf{v}'$ is $\mathbf{v}' = \mathbf{v}-\mathbf{u}$.

When we put $E = m_0c^2\gamma(v)$ and $p_x = m_0v_x\gamma(v)$ in Eq.(6.5) we get

$$p_x' = \gamma(u)[p_x-uE/c^2], \ p_y'=p_y, \ p_z'=p_z, \ E'=\gamma(u)[E-up_x], \tag{6.9}$$

and so, under a Lorentz transformation, the four components of $(\mathbf{p},E/c^2)$ transform like the four components of $(\mathbf{r},t)$. In the language of the next chapter $\mathbf{p},E/c^2$ forms an invariant 4-vector.

6.3 *The Limiting Velocity*

We have already seen that adding any two velocities, each less than c by however little, will always give a resultant velocity less than c. We now see another reason for the limiting effect of c. As v tends to c, the work required to accelerate a particle of finite rest mass, and the energy of the particle $E = m_0c^2/(1-v^2/c^2)^{1/2}$ both tend to infinity.

6.4 *The Zero Momentum Frame*

A set of interacting particles can be regarded as a single object with a definite total energy E, momentum $\mathbf{P}$ and inertial mass M and, for the system as a whole, $E=Mc^2$. Both E and M are constants. Since the total momentum $\mathbf{P}$ is also constant so is the quantity $E^2-P^2c^2$ which we might express as $(M_0c^2)^2$ regarding M_0 as the rest mass of the system made up from the rest masses of its constituent particles together with contributions from the internal kinetic and potential energies of the system. We could thus also write

$$Q^2 = E^2 - P^2c^2 \tag{6.10}$$

and regard Q as the energy available to be expended in terms of particle rest energies, potential energy and internal kinetic energy. This leads to the important result that, for a given <u>total</u> energy, Q is largest when $P = 0$.

If, in the laboratory frame two particles with a total energy E collide, the energy available to cause, for example, changes in their internal states, or the creation of new particles and radiation, is not E but $(E^2-P^2c^2)^{1/2}$. In other words the maximum <u>available</u> energy, i.e. the energy left over when the requirement that $\mathbf{P}$ be conserved has been met, is the energy calculated in a frame in which $P=0$. When two equal particles meet head on with equal velocities and energies, the laboratory frame is already the zero momentum frame and all the particle energy is available to cause reactions. If however one of the particles (the target particle) is stationary the available energy is less, and often much less, than the energy of the moving particle. A detailed calculation of the energy available in the zero momentum frame will be given in the next section.

The zero momentum frame is often called the centre of momentum frame, and also the centre of mass frame since it is the frame in which the position of the centre of mass is constant. We will call it the centre of momentum frame since it is the vanishing of the total momentum in this frame that is the crucial point. We will discuss the centre of mass frame in more detail in section (6.8).

6.5 *The Centre of Momentum Frame and Collisions*

Suppose that a particle of rest mass m_1, total energy E_1, momentum p_1 and velocity v_1, all in the laboratory frame F, collides with a stationary particle of rest mass m_2 and total energy $E_2=m_2c^2$. Initially the total momentum was p_1 and the total inertial mass $(E_1+E_2)/c^2$, thus the velocity of the centre of momentum frame was

$$u = p_1c^2/(E_1+E_2). \qquad (6.11)$$

In F, the velocity of the system as a whole was u and so, when we use Eq.(6.6a) to calculate E' in the centre of momentum frame, we must put $v = u$. This gives

$$E' = E\gamma(u)(1-u^2/c^2) = E(1-u^2/c^2)^{1/2} = E/\gamma(u). \qquad (6.12a)$$

With $E = E_1+E_2$ this becomes

$$E' = [(E_1+E_2)^2 - p_1^2c^2]^{1/2}. \qquad (6.12b)$$

Since $p_1^2c^2 = E_1^2-m_1^2c^4$ and $E_2 = m_2c^2$,

$$E' = c[m_1^2c^2 + m_2^2c^2 + 2m_2E_1]^{1/2} \qquad (6.12c)$$

or

$$E' = c[m_1^2c^2+m_2^2c^2+2m_2c(m_1^2c^2+p_1^2)^{1/2}]^{1/2}. \qquad (6.12d)$$

or

$$E'=c[m_1^2c^2+m_2^2c^2+2m_1m_2c^2\gamma(v_1)]^{1/2}, \qquad (6.12e)$$

and finally

$$E'=[m_1c^2+m_2c^2][1+2m_1m_2\{\gamma(v_1)-1\}/\{m_1+m_2\}^2]^{1/2}. \qquad (6.12f)$$

This is the most convenient form for discussing the classical limit for, as $v/c \to 0$, $\gamma(v)-1 \to \frac{1}{2}v^2/c^2$ and so

$$E'\to(m_1c^2+m_2c^2)[1+m_1m_2v_1^2/(m_1+m_2)^2c^2]^{1/2}, \qquad (6.13a)$$

and the available kinetic energy is

$$T'= \tfrac{1}{2}m_1m_2v_1^2/(m_1+m_2) = m_2T_1/(m_1+m_2) \qquad (6.13b)$$

where T_1 is the kinetic energy of the incident particle. This is, of course, the usual classical result. For example, if two equal particles collide, one being initially at rest, only half the incoming kinetic energy can be dissipated in the collision even if it is completely inelastic.

The extreme relativistic limit when E_1 is much greater than either m_1c^2 or m_2c^2 is best discussed using Eq.(6.12c) which yields $E' \rightarrow (2m_2c^2E_1)^{1/2}$, so that eventually the energy available in the centre of momentum system increases only as the square root of the incident energy. For example, if a proton with 100GeV energy collides with a stationary proton, the energy available to cause a reaction or the creation of a new particle is only 14GeV, whereas if two 50GeV protons make a head-on collision, the entire 100GeV is available. This is why elementary particle physicists desire colliding beam machines.

6.6 Elastic Scattering and the Compton Effect

We consider the elastic (energy conserving) scattering of a particle of rest mass m_o, momentum p_1 and energy e_1 by a stationary particle of mass M_o. The process is shown in Figure 6.2.

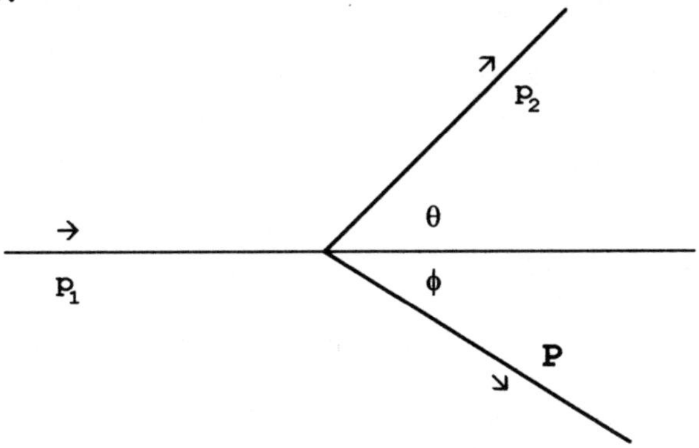

Figure 6.2 Elastic Scattering

From conservation of momentum we have $P\cos\phi = p_1 - p_2\cos\theta$, $P\sin\phi = p_2\sin\theta$, so that $P^2 = p_1{}^2 + p_2{}^2 - 2p_1p_2\cos\theta$. Conservation of energy yields $e_1 + E_1 = e_2 + E_2$, where $e_1{}^2 = p_1{}^2c^2 + m_o{}^2c^4$, $E_1{}^2 = M_o{}^2c^4$, $e_2{}^2 = p_2{}^2c^2 + m_o{}^2c^4$ and $E_2{}^2 = P^2c^2 + M_o{}^2c^4 = c^2(p_1{}^2 + p_2{}^2 - p_1p_2\cos\theta) + M_o{}^2c^4$. Thus $(e_1 - e_2 + M_oc^2)^2 = e_1{}^2 + e_2{}^2 - 2m_oc^4 - 2p_1p_2c^2\cos\theta + M_o{}^2c^4$. This gives

$$-e_1e_2 + M_oc^2(e_1 - e_2) = -m_oc^4 - p_1p_2c^2\cos\theta, \qquad (6.14a)$$

which can be re-arranged as

$$M_0 c^2(1/e_2-1/e_1) = 1-(m_0{}^2c^4+c^2p_1p_2\cos\theta)/e_1e_2. \quad (6.14b)$$

If, in (6.14a), we express p_2 in terms of e_2, then after a little algebra, we finally get e_2 in terms of e_1, p_1 and θ as

$$e_2 = [(e_1+M_0c^2)(m_o{}^2c^4+e_1M_oc^2) + c^2p_1{}^2\cos\theta(M_0{}^2c^4-m_0{}^2c^4\sin^2\theta)^{\frac{1}{2}}]$$
$$\div[(e_1+M_0c^2)^2-c^2p_1{}^2\cos^2\theta], \quad (6.15)$$

and then of course we can get p_2 from e_2.

A very considerable simplification occurs if the incident particle is a photon of zero rest mass $m_0 = 0$, for then $e_1 = p_1c$ and $e_2 = p_2c$, so that Eq.(6.14b) reduces to

$$M_0c^2(1/e_2 - 1/e_1) = 1 - \cos\theta \quad (6.16a)$$

or, in terms of the wavelengths λ_1 and λ_2, since $\lambda = hc/e$,

$$\lambda_2 - \lambda_1 = \lambda_0(1 - \cos\theta), \quad (6.16b)$$

where

$$\lambda_0 = h/cM_0. \quad (6.16c)$$

This is generally referred to as the Compton wavelength of the particle of rest mass M_0. For electrons λ_0=2.42×10^{-12}m and the fractional wavelength change is small unless the wavelength itself is also small, e.g. 10^{-11}m, corresponding roughly to 100keV Xrays. The precise agreement of the experimental Compton effect results with Eq.(6.16b) was one of the most important early verifications of quantum theory as well as of relativity.

6.7 Radiative Recoil

When an atom or a nucleus emits or absorbs radiation it will recoil because of the momentum of the radiation, and the kinetic energy associated with this recoil must be allowed for in the relation between the radiated energy and the change in the internal energy of the atom or nucleus.

Consider first a body at rest which emits electromagnetic radiation of energy Q and momentum Q/c in a definite direction while its internal energy changes from Mc^2+Q' to Mc^2. It recoils with a velocity v where $Mv\gamma(v)=Q/c$. The energy relation is $Mc^2+ Q' = Mc^2\gamma(v)+ Q$ and together these equations yield

$$\gamma(v) = [1+(Q/Mc^2)^2]^{\frac{1}{2}} \quad (6.17a)$$

and

$$Q' = Mc^2[(1+(Q/Mc^2)^2)^{\frac{1}{2}} -1] + Q. \quad (6.17b)$$

Next consider a body of mass M which absorbs radiation of energy Q and changes its internal energy by Q''. If the recoil velocity is v we have the momentum relation $(M+Q''/c^2)v\gamma(v)=Q/c$ and the energy relation $Mc^2+Q=(Mc^2+Q'')\gamma(v)$ so that

$$Q = Q''[1+(Q''/2Mc^2)] \tag{6.17c}$$

or

$$Q'' = Mc^2[(1+2Q/Mc^2)^{-1/2} -1]. \tag{6.17d}$$

So, for emission,

$$(Q'-Q)/Q' = \tfrac{1}{2}\, Q'/(Q'+Mc^2) \tag{6.17}$$

and for absorption

$$(Q-Q'')/Q'' = \tfrac{1}{2}\, Q''/Mc^2, \tag{6.19}$$

and in either case the right hand side of these equations is the ratio of the energy change to the initial energy of the body.

Suppose now that the radiation is a photon of angular frequency ω so that $Q = \hbar\omega$, while $Q' = Q'' = \hbar\Omega$ is the difference between two internal energy levels of an atom or nucleus then, for emission, we get $\omega/\Omega = 1-\hbar\Omega/2(\hbar\Omega + Mc^2)$ and for absorption, $\omega/\Omega = 1 + \hbar\Omega/2Mc^2$. For either γ-ray emission from nuclei or optical or X-ray emission from atoms $\hbar\Omega/Mc^2$ is small and so, in experiments to study resonant absorption by free nuclei or atoms of radiation emitted by identical atoms, the fractional detuning due to recoil will be $\Delta\omega/\omega \sim \hbar\omega/Mc^2$. For a proton $Mc^2 \sim 1$GeV, thus for atoms or nuclei Mc^2 will usually be a few tens of GeV. Since in optical spectra $\hbar\omega$ will be 10 eV or less, $\Delta\omega/\omega$ will be only 10^{-9} or so, less than the width of most spectral lines. In contrast, nuclear γ-rays with energies of 10keV to 1Mev have fractional natural widths which are very much smaller than the larger shift 10^{-6} to 10^{-4} due to recoil. As a result, whereas the experimental spectroscopy of atomic resonance absorption is almost unaffected by recoil, the corresponding study of nuclear radiation is only feasible because of the occurrence in some solids of recoil-less radiation. Here the nuclear recoil momentum is transmitted via the atomic electrons to the solid body as a whole, resulting in a negligible recoil velocity. This effect was first discussed theoretically by Lamb but is generally known as the Mossbauer effect, after its independent experimental discoverer.

We have already mentioned Pound and Snider's use of the Mossbauer effect in the first laboratory demonstration of the gravitational red shift predicted by general relativity.

It also has many applications in solid state physics and chemistry. Because the natural width $\Delta\omega/\omega$ of γ rays can be as small as 10^{-11}, very small changes in the observed γ ray energy due to slight differences in the environments of nuclei at different sites in a compound or crystal, or due to magnetic fields, can be detected. The Doppler shift is used to bring the emitting and absorbing nuclei back on resonance by giving them a relative motion, with either a mechanical system or an electromechanical transducer. We note that a relative velocity of only 3cm/sec. is needed to compensate for an energy shift of 1 part in 10^{10}, thus no very great demand is usually placed on the drive mechanism. On the other hand, complex and significant theoretical corrections are required to allow for the effect of atomic thermal vibrations. Although the linear effect of these vibrations averages out, the typical velocity v of thermal vibration, around 100m/s, gives a quadratic effect v^2/c^2 of about 10^{-13} which is not negligible. It corresponds to the gravitational red shift over a height of some 1000 metres (for further details see Pound and Rebka 1960).

If the quantum energy $\hbar\omega << Mc^2$, then the recoil velocity is given by $v/c \approx \hbar\omega/Mc^2$, but from Eq.(6.18) the frequency shift is $\Delta\omega/\omega = v/2c$. To this approximation it appears that what we are seeing is a Doppler shift associated with the average recoil velocity $v/2$ during the emission or absorption process. The effect is however essentially a *dynamical* effect involving the energies Q, Q' and Q''. Frequency only enters through the relation $Q = \hbar\omega$.

6.8 Centre of Mass Motion

When a system consists solely of a set of non-interacting particles of rest masses m_{on} and inertial masses $m_n = \gamma(v_n)m_{on}$ the total inertial mass is simply $M = \sum_n m_n$ and the centre of mass is at $\mathbf{R} = M^{-1}\sum m_n \mathbf{r}_n$. If, however, the particles interact the situation is rather more complicated, for we have to include the energy of interaction in the mass. Because this is continuously distributed throughout the system we will describe it by an energy density $\varepsilon(\mathbf{r})$ and a mass density ε/c^2. It will be convenient to include in $\varepsilon(\mathbf{r})$ the energy density associated with the particles themselves, in the sense that when a small volume element includes a particle of mass m the volume integral of ε over this volume contains a contribution mc^2 from the particle.

If a particle of inertial mass m has a velocity $\mathbf{v}$ we may think of it as transporting its energy $\varepsilon = mc^2$ at a velocity $\mathbf{v}$, or as constituting an energy flux $\mathbf{S} = \varepsilon\mathbf{v}$ (watts/m²). If we introduce a momentum density $\boldsymbol{\pi}$ we can also write $\mathbf{S} = c^2\boldsymbol{\pi}$. The rate at which energy crosses a plane element of area $d\mathbf{A}$ is $\mathbf{S}.d\mathbf{A} = c^2\boldsymbol{\pi}.d\mathbf{A}$. (We would have to be very careful to distinguish between the group and phase velocities if we were applying these ideas to waves described in terms of their frequency components.)

Consider now a small volume V bounded by a closed surface A with a positive outward normal. The rate at which energy leaves V is the surface integral $\iint \mathbf{S}.d\mathbf{A}$ and this must equal $-\iiint(\partial\varepsilon/\partial t)\,dV$, the rate at which the energy inside A is decreasing. Thus we have

$$\iint_A \mathbf{S}.d\mathbf{A} + \iiint_V(\partial\varepsilon/\partial t)\,dV = 0 \qquad (6.20a)$$

or, letting the volume $V\to 0$,

$$\mathrm{div}\mathbf{S} + \partial\varepsilon/\partial t = 0 \quad \text{or} \quad \boldsymbol{\nabla}.\mathbf{S} + \partial\varepsilon/\partial t = 0. \qquad (6.20b)$$

These both express energy conservation as typical continuity equations, and they can be used in conjunction with the relation

$$\mathbf{S} = c^2\mathbf{p}, \qquad (6.21)$$

to discuss the motion of the centre of mass located at

$$\mathbf{R} = [\iiint \mathbf{r}\varepsilon(\mathbf{r})\,dV]\,/\,[\iiint \varepsilon(\mathbf{r})\,dV], \qquad (6.22a)$$

where the integrals extend over the whole system out to a boundary in a region where both ε and $\mathbf{p}$ are zero. Since the denominator is just the total energy E of the system, which is constant, we have

$$\mathbf{R} = E^{-1}\iiint \mathbf{r}\,\varepsilon(\mathbf{r})\,dV(\mathbf{r}), \qquad (6.22b)$$

in which $\mathbf{r}$ is a variable of integration associated with the volume element dV and so, when we calculate $d\mathbf{R}/dt$, $\varepsilon(\mathbf{r})$ in the integrand only has to be differentiated with respect to t at fixed $\mathbf{r}$. Thus the velocity of the centre of mass is

$$d\mathbf{R}/dt = E^{-1}\int \mathbf{r}(\partial\varepsilon(\mathbf{r})/\partial t)\,dV = -E^{-1}\int \mathbf{r}\boldsymbol{\nabla}.\mathbf{S}\,dV. \qquad (6.23a)$$

The vector identity which, expressed in component form, is $\partial(r_i S_j)/\partial r_j \equiv S_i + (\mathbf{r}\boldsymbol{\nabla}.\mathbf{S})_i$, lets us express (6.23a) as

$$d\mathbf{R}/dt = -E^{-1}\int[\partial(r_i S_j)/\partial r_j]\,dV + E^{-1}\int \mathbf{S}\,dV.$$

The first volume integral yields an integral over a surface which is only restricted to the extent that it must include the whole system of particles and their interactions; thus it must be taken over a surface entirely outside the system, and there the integrand is zero. This leaves

$$d\mathbf{R}/dt = E^{-1}\int \mathbf{S}dV = E^{-1}c^2\int \mathbf{p}dV = M^{-1}\int \mathbf{p}dV = \boldsymbol{P}/M, \qquad (6.23b)$$

where $\boldsymbol{P}$ is the total momentum of the system. Since $\boldsymbol{P}$ is constant the centre of mass moves with a constant velocity. We see that, with the mass defined so that it includes the mass associated with the energy of interaction, we can retain the simple relation $\boldsymbol{P} = Md\mathbf{R}/dt$ between the momentum $\boldsymbol{P}$, the total inertial mass M (not just the rest mass) and the velocity $d\mathbf{R}/dt$ of the centre of mass. This important result allows us to separate the motion of a system of particles, or a composite body, into an internal motion superimposed on a steady motion of system as a whole; and then we can eliminate this steady motion from our calculations by an appropriate choice of a coordinate system.

As a simple example we consider a pulse of light passing at normal incidence through a plane-parallel slab of glass with a refractive index n and thickness L, as shown in the next figure. We ignore reflections which in any case could be eliminated using thin anti-reflection coatings. If the pulse energy is U and the slab, which was initially at rest,

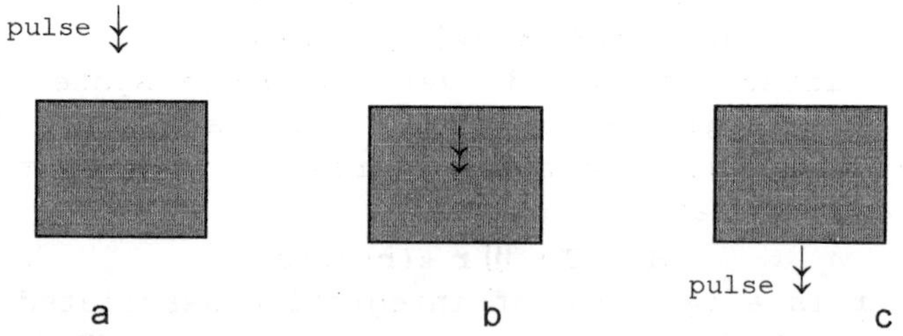

Figure 6.3 Centre of Mass Motion

has a mass M, then the centre of mass velocity in the initial state, in Figure 6.3a, is $V = cU/(U+Mc^2)$ and this will remain constant throughout the sequence of events shown in the three figures. Since $t=nL/c$ is the time taken by the pulse to traverse the slab, the centre of mass must have moved a distance Vt between Figures 6.3a and 6.3c but, if the slab does not move, the light pulse has moved only a distance L and so the slab itself must have been displaced a distance x given by

$$x = [Vt(U+Mc^2)-UL]/Mc^2 = (n-1)UL/Mc^2. \qquad (6.24a)$$

If ρ is the density of the slab and A is its cross-sectional area

$$x = (n-1)U/A\rho c^2. \qquad (6.24b)$$

A numerical example is given in problem 6.1 at the end of the chapter. You should find that the effect, though possibly measurable, is very small.

If, during the time t that the pulse is within the slab, the slab moves a distance x, its momentum must be Mx/t. The initial pulse momentum was U/c and, since momentum is conserved, the momentum

$$P_r = U/c-Mx/t = [1-(n-1)/n]U/c = U/cn \qquad (6.25a)$$

was not associated with the bulk motion of the slab. If this is expressed in terms of momentum $\boldsymbol{p}_r$ and energy u_r per unit volume then, since in the medium $u_r = n\mathbf{E}\times\mathbf{H}/c$ where $\mathbf{E}\times\mathbf{H}$ is Poynting's energy flux vector, we see that the momentum density is

$$\boldsymbol{p}_r = \mathbf{E}\times\mathbf{H}/c^2 \qquad (6.25b)$$

and the refractive index n has vanished from this relation, which has the same form as in vacuum. In vacuum p_r the momentum density of the wave can, of course, be derived directly from Maxwell's equations. The result (6.25a) is therefore often interpreted as meaning that the momentum density of a wave in a refracting medium is just $\mathbf{E}\times\mathbf{H}/c^2$. This is incorrect, all that we have shown is that this part of the momentum density is not associated with the motion of the slab as a whole. The separation of the momentum into a material part and a non-material part is not trivial. There are stresses in the medium in the vicinity of the wave and these correspond to momentum transfer occurring within the body.

Eq.(6.25a) suggests that the momentum of a photon would decrease as it enters a refracting medium. On the other hand, in quantum mechanics the momentum is hk ($hk \equiv h\nu/c$, in vacuum) and so might be expected to be $nh\nu/c$ in a medium, and this is greater than in vacuum. The subject has been discussed almost continuously since the early years of the century starting with Abraham and Minkowski, and controversy has persisted until the present day involving a succession of distinguished physicists, some supporting one view and others the opposite. In fact they have all been wrong for as Shockley (1968) showed, in a short and ingenious paper, questions about the momentum of photons in a refracting medium cannot be answered in terms of the refractive index alone; other more subtle properties of the medium are involved. Robinson (1975) has shown that Shockley's conclusions are in complete agreement with a treatment of stress given by Helmholtz, many years before the discovery of either relativity or quantum mechanics.

The history of this rather pointless controversy serves once again as a warning against the danger of appealing to intuition or that popular oxymoron, the thought experiment, as a substitute for careful analysis and calculation.

6.9 Force

There are two notions implicit in the concept of force. The first is the notion of a force $\mathbf{f}$ causing momentum to change at a rate $d\mathbf{p}/dt = \mathbf{f}$. The second is that forces are caused by the present, and possibly the past, configuration of other particles. If we symbolise this configuration as C, we can express the overall relation by

$$d\mathbf{p}/dt = \mathbf{f}(C) \tag{6.26}$$

In electromagnetism, for example, we have $d\mathbf{p}/dt = q(\mathbf{E} + \mathbf{v}\times\mathbf{B})$ in terms of the particle's charge q and velocity $\mathbf{v}$, and two fields $\mathbf{E}$ and $\mathbf{B}$ generated by other charged bodies.

If this is to be useful in relativity both the relation of force to momentum and to its sources must be Lorentz invariant, i.e. independent of any particular choice of an inertial coordinate system.

Taking the relative velocity of two frames F' and F along the x axis as usual, and considering a particle of velocity $\mathbf{v}$ relative to F, we have $t' = \gamma(u)(t-ux/c^2) = t\gamma(u)(1-\mathbf{u.v}/c^2)$ and the transformation laws for the components of $\mathbf{p}$ are

$$p_x' = \gamma(u)[p_x - um_0\gamma(v)] = \gamma(u)[p_x - eu/c^2] \tag{6.27a}$$

$$p_y' = p_y, \qquad p_z' = p_z. \tag{6.27b,c}$$

We note that $d\mathbf{p}/dt = \mathbf{f}$ and that $de/dt = \mathbf{v.f}$, so that

$$\mathbf{f}' = d\mathbf{p}'/dt' = (d\mathbf{p}'/dt)/(dt'/dt),$$

which gives

$$f_x' = [f_x - u(\mathbf{f}.\mathbf{v})/c^2]/(1 - \mathbf{u}.\mathbf{v}/c^2)], \qquad (6.28a)$$

$$f_y' = f_y/[\gamma(u)(1 - \mathbf{u}.\mathbf{v}/c^2)], \qquad (6.28b)$$

$$f_z' = f_z/[\gamma(u)(1 - \mathbf{u}.\mathbf{v}/c^2)]. \qquad (6.28c)$$

We see that the transformation law for force depends not only on u, but also on the velocity v of the particle on which the force is acting. If the force happens, like the electrostatic force, to be independent of velocity in F, it may become velocity dependent in F'. Moreover if in F we have carefully separated the force into a part dependent on the particle alone and a part dependent on a field, this separation need not have the same form in F'. This suggests that velocity independent forces such as classical gravity or electrostatics are unlikely to be Lorentz invariant and will turn out, in relativity, to be part of some more general scheme. The complexity of the force transformation also suggests that the traditional classical form of force is unlikely to remain such a useful simplifying concept.

6.10 Forced Particle Motions

We now look at a few basic particle motions in impressed, or given, fields. We begin with the equation of motion

$$d\mathbf{p}/dt = \mathbf{f}(\mathbf{r}, t, \text{etc.}) , \qquad (6.29)$$

where the prescribed force $\mathbf{f}$ is a given function of the position $\mathbf{r}$ of the particle, the time t and possibly the velocity $\mathbf{v}$ of the particle as well. We have $\mathbf{p} = m\gamma(v)\mathbf{v}$ (where m is now the <u>rest</u> mass of the particle) and $d\mathbf{r}/dt = \mathbf{v}$, so that (6.29) becomes

$$md(\mathbf{v}\gamma(v))/dt = md/dt[\{1-(dr/dt/c)^2\}^{-\frac{1}{2}}d\mathbf{r}/dt] = \mathbf{f}(\mathbf{r}, t, \text{etc}).$$

This is appreciably more complicated than the classical equation $md^2\mathbf{r}/dt^2 = \mathbf{f}$, but, because it is still a second order differential equation, it will determine a unique trajectory if $\mathbf{r}$ and $d\mathbf{r}/dt$ are given at some initial time. Like the classical equation it only has simple solutions for a few problems, such as the harmonic oscillator and uniformly accelerated motion. Most other cases have to be treated by approximation, or perturbation, methods. Here we confine ourselves to a few of the simplest cases and, in particular, to cases where the force is independent of the time.

If the force is conservative so that $\mathbf{f} = -\mathbf{grad}\Phi$, then the total energy E_0 is constant and, if m is the rest mass of the particle, we have $mc^2\gamma(v) = E_0 - \Phi$, which gives

$$(E_0 - \Phi)(1 - v^2/c^2)^{1/2} = mc^2. \qquad (6.30a)$$

This can be solved for the magnitude v of the velocity and then, at least in one dimensional problems, one more integration gives x. Even this is rarely straightforward unless the form of $\Phi(x)$ is particularly simple, but then it would hardly be much simpler in classical mechanics.

In three dimensional problems we can use Eq.(6.30a) to eliminate $\gamma(v)$ and thus obtain

$$d/dt[\mathbf{v}(E_0 - \Phi)/c^2] = (\mathbf{v}.\mathbf{grad})[\mathbf{v}(E_0 - \Phi(\mathbf{r}))/c^2 = -\mathbf{grad}\Phi. \qquad (6.30b)$$

Very few forms of the potential Φ yield simple solutions of this equation. One of the more important cases is circular motion in a central field, for then $|\mathrm{v}|$ is constant so that

$$m\gamma(v)\,d\mathbf{v}/dt = -\mathbf{grad}\Phi, \qquad (6.30c)$$

which, apart from the <u>constant</u> factor $\gamma(v)$, is the same as the classical equation.

In the case of a particle in a harmonic potential well where $\Phi = \frac{1}{2}kr^2$ the angular frequency of <u>circular</u> motion is $\omega = [k/m\gamma(v)]^{1/2}$. Because $v = \omega r$ the angular frequency will vary with r. If we set $\omega_0 = (k/m)^{1/2}$ we obtain

$$(\omega/\omega_0)^2 = [1 + \tfrac{1}{2}(r\omega_0/c)^4]^{1/2} - \tfrac{1}{2}(r\omega_0/c)^2.$$

When $r\omega_0 << c$ this gives

$$\omega \approx \omega_0[1 - \tfrac{1}{2}(r\omega_0/c)^2]. \qquad (6.30d)$$

The frequency therefore decreases as r increases.

A charged particle in a constant and uniform magnetic field $\mathbf{B}$ experiences a force at right angles to its motion, and so its energy and its speed v remain constant. The equation of motion becomes $m\gamma(v)\,d\mathbf{v}/dt = q\mathbf{v}\times\mathbf{B}$ giving circular motion about the direction of $\mathbf{B}$ with angular frequency

$$\omega = qB/\{m\gamma(v)\}. \qquad (6.30e)$$

This gives the relation between frequency ω and particle energy $m\gamma(v)$, and is a useful relation in the design of cyclic particle accelerators. Since $v = \omega r$ we can derive a variety of other results, for example,

$$r = mv\gamma(v)/qB = p/qB, \qquad (6.30f)$$

so that r is simply proportional to the momentum. This is especially useful in analysing cloud chamber and bubble chamber tracks of high energy particles. For very high energy particles with $E >> mc^2$, we have $E \sim pc$ and so $r \sim E/cqB$, thus with 30 Gev protons and B=1T, it would need r=100m and a magnet over 150m long to bend a beam through 90°.

As a last example we consider the one dimensional problem of a body, of rest mass m, subject to a constant force F, for example an electron in a constant electric field. The force corresponds to a potential $\Phi = -Fx$ and, if $x = 0$ and $dx/dt = 0$ at $t = 0$, we have $mc^2\gamma(v) = mc^2+xF$, which gives the relation $(dx/dt)^2 = c^2[1-\{mc^2/(mc^2+Fx)\}^2]$ so that $cdt = (mc^2+Fx)dx/[(mc^2+Fx)^2-m^2c^4]^{1/2}$. This is readily integrated to give $ct = (x^2+2xmc^2/F)^{1/2}$, and from this we get

$$x = [mc^2/F][(1 + F^2t^2/m^2c^2)^{1/2} - 1]. \tag{6.31}$$

For $Ft \ll mc$ this gives the classical result $x = \frac{1}{2}Ft^2/m$, and for $Ft \gg mc$, $x \rightarrow ct$ as we might expect. If we identify F/m with the constant acceleration a, it is identical with Eq.(3.10f), $x = (c^2/a)[(1+a^2t^2/c^2)^{1/2} -1]$ and this confirms the idea that we introduced in Chapter 3: that constant acceleration in the instantaneous, comoving, rest frame of a particle corresponds to a constant force.

6.11 Summary

We started our development of special relativity by adding the universal constancy of the velocity of light to a few rather general assumptions about the nature of space and time. The resulting Lorentz transformation specifies the relation between the coordinates in time and space ascribed to the same event, when we use two different inertial systems in uniform relative motion.

We then explored how this new relation modifies the kinematic description of physical phenomena when speeds comparable with c are involved, the most obvious effects being time dilation and the relativistic Doppler effect. Next, in Chapter 5, we looked at the consequences of requiring conservation of linear momentum in a simple collision to be valid in any inertial frame, if it were valid in one particular frame. This led to replacing $\mathbf{p}=m\mathbf{v}$ by $\mathbf{p}=m\gamma(v)\mathbf{v}$, and thus to new relations such as $E^2=p^2c^2+m^2c^4$ and the new form of dynamics used in this chapter. A theory which started by describing how descriptions of events in two different frames are related has thus generated new dynamical laws operating within a single reference frame.

The behaviour of an equation expressing a physical law, when the coordinate system is subjected to a Lorentz transformation, determines whether that law and the principles of special relativity are compatible.

A relativistically valid law retains the same form after a Lorentz transformation of the coordinate system.

Problems

6.1 A high pulse energy, even for a laser, at a wavelength of around 10^{-6} m, is about 10^4J, so that the largest reasonable value for U/A in Eq. (6.24b) is about 10^{15}J/m². Assume that ρ = 1000kg/m³ (the density of water) and n = 1.33 and calculate the displacement of the block in Figure 6.3.

6.2 A spherical vessel of radius 1m contains 10^{25} helium atoms at 1000K. What fraction of the pressure exerted on the vessel is due to radiation?

6.3 A 100keV photon is scattered through 90 degrees by an electron. What is the recoil velocity of the electron?

6.4 Show by considering conservation of energy and momentum that electron-positron annihilation can only proceed by the emission of two photons.

6.5 A proton of mass M and total energy $E=2Mc^2$ is scattered through 90 degrees by a stationary alpha particle of mass $4M$. Find the energy of the scattered proton.

6.6 A particle of rest mass m moving along the x axis, in response to a constant force F, has $x=0$, $v=dx/dt=0$ at $t=0$. Integrate the equation of motion $m d/dt[v(1-v^2/c^2)^{-1/2}]=F$ to obtain v and x as functions of t, and compare your results with equation(6.31).

6.7 A photon of energy E collides head on with a photon of energy 4×10^{-23}J, corresponding to the cosmic black body radiation at 3K. Show that electron-positron pair creation is possible if E exceeds about 10^{15} eV.

CHAPTER 7

4-VECTORS

We now need a better and more compact notation for dealing with relativistic equations, something analogous to the use of vectors in classical physics. Although vector notation improves the appearance of the equations of classical physics, when we want numerical results we still have to revert to the use of a coordinate system. In relativity we are much more closely tied to the use of coordinate systems but we can still introduce a notation which improves the appearance of relativistic equations and, in the conventional jargon, makes their invariance under Lorentz transformations manifest.

The fundamental invariant relativistic concept corresponding to distance is the interval between two events. It has a fourth time component as well as its usual three space components. We shall therefore be dealing with entities whose components come in sets of *four* rather than *three*, with one member of the set being different in kind from the other three. We cannot therefore generalise the coordinate independent form of vector analysis in which directed quantities are denoted by single bold faced symbols. The rules for manipulating these symbols without recourse to a coordinate system, e.g. vector addition and the two types of product, are based on geometrical rules which no longer apply. There are **no** *geometric* rules for combining time displacements with space displacements! We need a scheme with rules that are entirely *algebraic*. In ordinary vector analysis this is accomplished by treating a vector **R** as an ordered set of three numbers, its components R_j, with j = 1,2,3. Geometric operations on **R** are then replaced by algebraic operations on the three numbers R_j. Thus instead of $\mathbf{R.S} = RS\cos\angle RS$ we have $\Sigma_j R_j S_j$. Although the algebraic structure clearly has a geometric application, it is itself entirely independent of geometry. It can, therefore, be generalised to deal with ordered sets of *four* numbers F_j where j now runs from 1 to 4.

7.1 4-Vectors Defined

The basic concept in special relativity which replaces the spatial displacement **r** is the separation, or interval, between two events; this has a spatial part **r** and a time part t. In relativity we require all physically meaningful equations to be invariant under Lorentz transformations as well as rotations or translations of the spatial axes. The Lorentz transformation does not leave either r^2 or t^2 invariant, but it does leave the interval

$$s^2 = r^2 - c^2t^2 \tag{7.1}$$

invariant and, since under rotations of the axes both terms in this expression are separately invariant, (7.1) is invariant under both rotations and Lorentz transformations to new inertial coordinate systems. The negative sign in (7.1) can be dealt with in a number of ways but the simplest is Minkowski's notation which introduces a fourth component r_4 or $x_4 = ict$ (where i is $\sqrt{-1}$). The invariant interval is then $r_k r_k$ where now, and in every similar expression, a repeated subscript is to be summed from 1 to 4, although an equation such as $v_j = dx_j/d\tau$ stands for four separate equations as j takes on the values 1,2,3,4 in succession.

The disadvantage of this notation is that it pre-empts $i=\sqrt{-1}$, which is no longer available for dealing with waves and oscillations, or in wave mechanics. This can be avoided with contravariant and covariant vectors $x_j \to (x_o = ct, x_1, x_2, x_3)$ $x^j \to (x_o, -x_1, -x_2, -x_3)$, and a metric tensor $g^j_k = \mathrm{diag}(1,-1,-1,-1)$, so that the invariant interval becomes $x_j x^j = x_j g^j_k x_k$. For our purposes this is hardly worthwhile.

A linear transformation of the coordinates of an event such as a Lorentz transformation or a rotation of the axes can be expressed in terms of a matrix whose elements a_{jk} are the coefficients in the transformation equations. When

$$x_k \to x_j' = a_{jk}x_k \tag{7.2a}$$

(remember this is summed over k) the matrices a_{jk} will only leave Eq.(7.1) invariant if $x_j'x_j' = a_{jk}a_{jl}x_k x_l$ is identical with $x_k x_k$ for all x_k and so the matrix elements must satisfy

$$a_{jk}a_{jl} = \delta_{kl}, \quad a^{-1}{}_{jk} = a_{kj}, \quad a_{ik}a_{jk} = \delta_{ij}, \quad |a| = \pm 1. \tag{7.2b}$$

This allows Lorentz transformations, rotations, reflections and inversions. The "improper" reflections and inversions are eliminated by insisting on a positive determinant, i.e.

$$|a| = +1. \tag{7.2c}$$

Permissible relativistic transformations may also include translations of the origin as well as products of rotation matrices Θ_{jk} and the special Lorentz transformation, or *boost*, B_{jk}, for which

$$r_1' = \gamma(u)(r_1 - ut),\ r_2' = r_2,\ r_3' = r_3,\ t' = \gamma(u)(t - ur_1/c^2). \tag{7.3}$$

If we let $\beta = u/c$, $\gamma(\beta) = (1-\beta^2)^{-1/2}$, and replace r_1, r_2, r_3, t by $x_1, x_2, x_3, x_4 = ict$ this transformation becomes

$$x_1' = \gamma(\beta)(x_1 + i\beta x_4),\ x_2' = x_2,\ x_3' = x_3,\ x_4' = \gamma(\beta)(x_4 - i\beta x_1), \tag{7.4}$$

and its matrix is

$$B_{jk} = \begin{matrix} \gamma & 0 & 0 & i\beta\gamma \\ 0 & 1 & 0 & 0 \\ 0 & 0 & 1 & 0 \\ -i\beta\gamma & 0 & 0 & \gamma\ . \end{matrix} \tag{7.5}$$

It is easy to check that this satisfies Eq.(7.2b).

A rotation matrix in 4-component notation is formed by augmenting an ordinary 3×3 rotation matrix with 1 in the fourth row and column, thus

$$\Theta(4\times4) = \begin{matrix} \Theta(3\times3) & 0 \\ 0 & 1\ . \end{matrix}$$

Any quantity Y with 4 components Y_1, Y_2, Y_3, Y_4 which transform like x_1, x_2, x_3, x_4 is a **covariant 4-vector**.

7.2 Some Simple 4-Vectors

We begin with the relativistic velocity which generalises $\mathbf{v} = d\mathbf{r}/dt$. Obviously we add the fourth component $r_4 = ict$ to complete the 4-vector r_j , but we cannot consider dr_j/dt as a suitable generalisation since dt is not a scalar invariant but depends on the particular coordinate system. We have available, however, a suitable scalar invariant, which reduces to t when only velocities small compared with c are involved. This is the <u>proper time</u> τ defined by

$$d\tau = [(dt)^2 - d\mathbf{r}.d\mathbf{r}/c^2)]^{1/2} \tag{7.6}$$

and because $d\mathbf{r} = \mathbf{v}dt$, we also have $d\tau = dt/\gamma(v)$. The quantity

$$V_j = dx_j/d\tau \tag{7.7}$$

is thus a covariant 4-vector. Its 4 components can be expressed as

$$V_j = \gamma(v)[dx_1/dt, dx_2/dt, dx_3/dt, ic] \tag{7.8}$$

and when $v \ll c$ it reduces to $V_j = (v_1, v_2, v_3, ic)$. The first three components are just the classical velocity and the fourth is a harmless and uninteresting constant with no particular classical interpretation. The method by which we arrived at V_j as the relativistic generalisation of the classical velocity **v** is typical. We took the classical relation $\mathbf{v} = d\mathbf{r}/dt$, replaced **r** by its obvious generalisation r_j, rejected dt in the denominator as not covariant and replaced it by the invariant proper time $d\tau$ which reduces to dt as $v/c \to 0$.

Next we consider the relativistic generalisation of the relation between momentum and velocity i.e. $\mathbf{p} = m\mathbf{v}$. A very obvious candidate is

$$p_j = m_o V_j \tag{7.9}$$

where m_o is the invariant rest mass. Because V_j is a 4-vector so is p_j. The components of p_j are

$$p_j = \gamma(v) m_o (v_1, v_2, v_3, ic) = \gamma(v) m_o (v_1, v_2, v_3), iE/c \tag{7.10}$$

where $E=m_o c^2 \gamma(v)$ is the energy. The first three components are already familiar as the components of the relativistic momentum introduced in Chapter 5, and they reduce to the classical momentum as $v/c \to 0$. The fourth component reduces, as $v/c \to 0$, to a constant $im_o c$ which does not appear in classical equations. We have an important new result.

'The three components of the relativistic momentum and the energy together form a 4-vector'.

The associated scalar invariant is $p_j p_j = \mathbf{p}.\mathbf{p} - E^2/c^2$ but from (7.10) it is also $\gamma(v)^2 m_o^2 (v_1^2+v_2^2+v_3^2-c^2) = -m_o^2 c^2$ and we thus recover the basic dynamical equation $E^2=p^2c^2+m_o^2c^4$, which can now be written in the manifestly covariant form

$$p_j p_j = - m_o^2 c^2 . \tag{7.11}$$

The obvious covariant generalisation of $\mathbf{F}=d\mathbf{p}/dt=md\mathbf{v}/dt$ is

$$F_j = dp_j/d\tau = m_o dV_j/d\tau . \tag{7.12}$$

In the classical limit, $v \ll c$, the first three components reduce to the usual classical result, but the last component is $F_4 = m_o d[ic\gamma(v)]/d\tau = (i/c)dE/d\tau$ and, in the classical limit, $-icF_4$ will become simply dE/dt, the rate at which the force is doing work to increase the energy of the particle. Since $dE/d\tau = \gamma(v)dE/dt$ we can also write $-icF_4$ as $\gamma(v)\mathbf{F}.\mathbf{v}$ in terms of the classical force and velocity. Thus another expression for the 4-force is

$$F_j = \gamma(v)(\mathbf{F}, i\mathbf{F}.\mathbf{v}/c) . \tag{7.13}$$

From Eq.(7.8) and Eq.(7.13) the invariant scalar $F_j V_j$ is

$$F_j V_j = \gamma^2(\mathbf{v.F} - \mathbf{v.F}) = 0 \qquad (7.14)$$

which is very different from the classical value of **F.v.** If, however, we express F_j as $dp_j/d\tau = \gamma(v)dp_j/dt$ with $p_4 = iE/c$, Eq. (7.14) gives the relation

$$\gamma^2(v)[v_1dp_1/dt + v_2dp_2/dt + v_3dp_3/dt - dE/dt] = 0$$

in which **v** is the ordinary velocity but p_1, p_2, p_3 and E are the relativistic momentum and energy. Since γ is non-zero this yields the relation

$$dE = v_1dp_1 + v_2dp_2 + v_3dp_3, \qquad (7.15a)$$

or

$$v_\lambda = \partial E/\partial p_\lambda \qquad (7.15b)$$

where the Greek subscript λ = 1,2 or 3 but *not* 4. We adopt this common convention that whereas Roman subscripts run from 1 to 4, Greek subscripts only run from 1 to 3.

The 4-force can be written as

$$F_j = m dV_j/dt = [m/\gamma(v)]dV_j/d\tau = m_0 dV_j/d\tau \qquad (7.16)$$

and the first form $m dV_j/dt$ is not *manifestly* covariant as neither m nor dt are covariant scalars but it is in fact, as we have just seen, covariant. This introduces a new idea. We can sometimes combine two non-covariant quantities such as m and dt to give a covariant product or quotient.

An obvious generalisation of the acceleration $\mathbf{a} = d\mathbf{v}/dt$ is

$$a_j = dV_j/d\tau = \gamma(v)d(\gamma\mathbf{v}, ic\gamma)/dt \qquad (7.17)$$

and since

$$d\gamma/dt = d(1-v^2/c^2)^{-1/2}/dt = \mathbf{v.a}\ \gamma(v)^3/c^2 \qquad (7.18)$$

we have

$$a_j = \gamma^2(v)[c^{-2}(\mathbf{v.a})\mathbf{v}\gamma^2(v) + \mathbf{a},\ i\gamma^2\mathbf{v.a}/c]. \qquad (7.19)$$

When $v \ll c$ this, to first order in v/c, reduces to

$$a_j \rightarrow [\mathbf{a}\ , i\mathbf{v.a}/c]. \qquad (7.20)$$

We notice how complicated the expression for a_j is in terms of **a** and **v** and also that the fourth component is of *first* order in v/c.

The scalar invariant $V_j a_j$ is

$$\gamma^3[\gamma^2\mathbf{v.a}\ v^2c^{-2} + \mathbf{v.a} - \gamma^2\mathbf{v.a}] = \gamma^3[\mathbf{v.a} - \gamma^2(1-v^2c^{-2})\mathbf{v.a}],$$

and, since $\gamma^2(1-v^2c^{-2}) = 1$, we obtain

$$V_j a_j = 0. \qquad (7.21)$$

This is obviously related to the result $V_j F_j = 0$, since clearly we have

$$F_j = m_0\ a_j. \qquad (7.22)$$

7.3 Vector Products and Tensors

In 3-dimensional vector analysis the algebraic form of the equation $\mathbf{H} = \mathbf{F}\times\mathbf{G}$ is $H_i = \varepsilon_{ijk}F_jG_k$ and we might expect something similar in 4 dimensions involving the totally antisymmetric symbol with subscripts running from 1 to 4. The only such symbol is ε_{pqrs} with 4 subscripts, which is zero except when pqrs is an even permutation of 1234, when it is +1, or when pqrs is an odd permutation of 1234, when it is -1. This is the symbol used in the expansion of a 4×4 determinant and, if **a** is the matrix of a Lorentz transformation,

$$\varepsilon_{pqrs}a_{pi}a_{qj}a_{rk}a_{sl} = \pm\det\mathbf{a} = \pm 1 \qquad (7.23)$$

where + applies if ijkl is an even permutation of 1234 and - if it is odd. If two or more subscripts are equal the result is zero and so ε is an invariant for

$$\varepsilon_{pqrs}\, a_{pi}a_{qj}a_{rk}a_{sl} = \varepsilon_{ijkl}. \qquad (7.24)$$

Consider then the nearest that we can get to generalising a vector product, i.e. $\varepsilon_{ijkl}F_kG_l$. This is clearly not a 4-vector, it has two running subscripts, i and j. Its transformation law is $\varepsilon_{ijkl}F_kG_l \rightarrow \varepsilon_{ijkl}F_k'G_l' = \varepsilon_{ijkl}a_{kr}a_{ls}F_rG_s$. Multiply both sides of this equation by $a_{ip}a_{jq}$ and use Eq.(7.24); this gives $a_{ip}a_{jq}\varepsilon_{ijkl}F_k'G_l' = \varepsilon_{pqrs}F_rG_s$; now multiply again by $a_{gp}a_{hq}$ and use $a_{gp}a_{ip} = \delta_{gi}$ and $a_{hq}a_{jq} = \delta_{hj}$. This finally gives $\varepsilon_{ghkl}F_k'G_l' = a_{gp}a_{hq}\varepsilon_{pqrs}F_rG_s$ and so the quantity transforms like the product of two 4-vectors. It is a covariant tensor of rank two.

In general, a covariant tensor of rank n is a quantity with n subscripts that transforms like the product of n 4-vectors. In three dimensions some typical 2nd rank tensors are strain and stress, others are dielectric constants in anisotropic media, where $D_\lambda = \varepsilon_{\lambda\mu}E_\mu$ with a sum over $\mu = 1,2,3$. There are also 3rd rank tensors such as the piezoelectric coefficients, relating strain to electric field and 4th rank tensors such as the electrostrictive coefficients or the elastic moduli relating stress to strain. In relativity the most important 2nd rank tensors are associated with angular momentum and the curl of a vector which, in 3 dimensions, are more concisely expressed by a vector product.

There is a very simple reason why the equivalent of the vector product in four dimensions is not just a 4-vector.

In three dimensions two vectors **A** and **B** define a plane and this defines a unique direction (its positive normal) which can be used to specify the direction of **A**×**B**. Two 4-vectors are insufficient to define a third unique direction in four dimensions.

There is, however, another and much more compact way of dealing with the analogue of the vector product of two 4-vectors F_i and G_i. This is to consider instead the 2nd rank, antisymmetric, covariant tensor whose components are given by

$$H_{ij} = F_iG_j - F_jG_i. \qquad (7.25)$$

Its diagonal elements with j=i are obviously zero and, for example, one of its off-diagonal elements is $H_{12} = F_1G_2 - F_2G_1$ which is the third component of **F**×**G**. The 6 non-zero components of H_{ij} with purely spatial subscripts i,j = 1,2,3 simply repeat the three components of **F**×**G** twice, once with each sign. This 2nd rank tensor is the simplest way we can incorporate the components of a vector product within a co-variant symbol. We will eventually see that the mixed space time components also have their uses.

In classical mechanics the angular momentum of a particle is **r**×**p**. A covariant generalisation of this, in terms of x_j and the relativistic momentum $p_j = (\mathbf{p}, iE/c)$, is

$$L_{ij} = x_ip_j - x_jp_i \qquad (7.26a)$$

whose space-like part repeats the components of the vector product twice. The classical torque, or couple, is $\mathbf{\Gamma} = \mathbf{r}\times\mathbf{F}$ with a relativistic generalisation $\Gamma_{ij} = x_iF_j - x_jF_i$.

The equation $d\mathbf{L}/dt = \mathbf{\Gamma}$ becomes

$$dL_{ij}/d\tau = \Gamma_{ij}. \qquad (7.26b)$$

and, in the classical limit, its space components repeat the classical equation twice over. Their relativistic form contains terms such as $d[r_1mv_2 - r_2mv_1]/dt = r_1F_2 - r_2F_1$ and since $F_2 = d(mv_2)/dt$ and $dr_1/dt = v_1$ etc. this is clearly a valid relation. The mixed space-time components yield $d(m\mathbf{r} - tm\mathbf{v})/dt = \mathbf{r}(\mathbf{F}.\mathbf{v})/c^2 - t\mathbf{F}$ and now, since $d(m\mathbf{v})/dt = \mathbf{F}$ and $dm/dt = c^{-2}dE/dt = c^{-2}\mathbf{F}.\mathbf{v}$, this is also a valid relation.

The extra components added to the components of **r**×**p** are useful in other contexts. Thus in systems of interacting particles they describe the motion of the centre of mass, and as components of the generalised curl they play an important role in formulating electromagnetism in 4-vector notation.

7.4 *Partial Derivatives*

If ϕ is a scalar the generalisation of $\mathbf{grad}\phi \equiv \nabla\phi$ with components $\partial\phi/\partial x_1$, $\partial\phi/\partial x_2$, $\partial\phi/\partial x_3$, is $\partial\phi/\partial x_j$ which gives $d\phi = \Sigma_j(\partial\phi/\partial x_j)dx_j$. Since $d\phi$ is a scalar $d\phi' = d\phi$ and therefore $\partial\phi/\partial x_j' a_{jk} dx_k = \partial\phi/\partial x_k dx_k$. Thus $a_{jk}\partial\phi/\partial x_j' = \partial\phi/\partial x_k$ must be valid for arbitrary dx_k. We multiply this equation on both sides by a_{ik} and use $a_{ik}a_{jk} = \delta_{ij}$ to obtain $\partial\phi/\partial x_i' = a_{ik}\partial\phi/\partial x_k$ so that

$$\partial/\partial x_i' = a_{ik}\,\partial/\partial x_k, \tag{7.27}$$

which is the usual transformation law for a 4-vector.

If F_j is a 4-vector then $\partial F_j/\partial x_j$ is a scalar and is the generalisation of $\mathrm{div}\mathbf{F}$ or $\nabla.\mathbf{F}$, while the quantity

$$C_{ij} = \partial F_j/\partial x_i - \partial F_i/\partial x_j \tag{7.28}$$

is an antisymmetric covariant 2nd rank tensor, the natural generalisation of **curlF**. Its mixed space-time components give $\nabla F_4 - (1/ic)\partial\mathbf{F}/\partial t$ and so C_{ij} contains derivatives other than **curl** associated with F_j. We shall see in Chapter 8 that if $F_j = (\mathbf{A}, i\phi/c)$, where $\mathbf{A}$ and ϕ are the vector and scalar electromagnetic potentials, then C_{ij} in Eq.(7.28) is a tensor constructed from the six components of the electric and magnetic fields $\mathbf{E}$ and $\mathbf{B}$. We can obviously construct further covariant derivatives, for example, if F_j is a covariant vector and L_{ij} is a covariant tensor, $\partial^2 F_j/\partial x_k \partial x_l$ and $\partial L_{ij}/\partial x_k$ are both 3rd rank covariant tensors.

The generalisation of the Laplacian operator ∇^2 is the d'Alembertian

$$\Box^2 = \nabla^2 - c^{-2}\partial^2/\partial t^2 = \partial/\partial x_j \partial/\partial x_j, \tag{7.29}$$

clearly an invariant scalar operator. This has the obvious and important consequence that, if Ψ is a scalar, the wave equation $\Box^2\Psi \equiv \nabla^2\Psi - c^{-2}\partial^2\Psi/\partial t^2 = 0$ is an invariant equation.

7.5 *Symmetric and Antisymmetric Tensors*

Many of the most important second rank tensors are either symmetric with $T_{jk} = T_{kj}$ or antisymmetric with $T_{jk} = -T_{kj}$. In any case any tensor T_{jk} can be written in terms of its symmetric part, $S_{jk} = \frac{1}{2}(T_{jk}+T_{kj})$, and its antisymmetric part, $A_{jk} = \frac{1}{2}(T_{jk} - T_{kj})$, as $T_{jk} = S_{jk} + A_{jk}$.

Under a Lorentz transformation $T_{jk}' = a_{jm}a_{kn}T_{mn}$ and so, if $T_{mn} = T_{nm}$ then $T_{jk}' = T_{kj}'$, and if $T_{mn} = -T_{nm}$ then $T_{jk}' = -T_{kj}'$. Thus symmetry and antisymmetry are Lorentz invariant properties. The most general 2nd rank tensor has 16 independent components, its symmetric part has 10 and its antisymmetric part has only 6, which is the number of components of two ordinary 3-vectors. We now show that <u>any</u> antisymmetric tensor A_{jk} is associated with two three-dimensional vectors. Consider the effect of a rotation through an angle θ about the 3-axis described by a matrix whose only non-zero elements are $a_{11}=a_{22}=\cos\theta$, $-a_{12}=a_{21}=\sin\theta$, $a_{33}=a_{44}=1$; this gives

$$A_{23}' = a_{22}a_{33}A_{23}+a_{21}a_{33}A_{13} = A_{23}\cos\theta - A_{31}\sin\theta,$$
$$A_{31}' = A_{31}\cos\theta + A_{23}\sin\theta, \qquad A_{12}' = A_{12},$$
$$A_{41}' = A_{41}\cos\theta - A_{42}\sin\theta,$$
$$A_{42}' = A_{42}\cos\theta + A_{41}\sin\theta, \qquad A'_{43} = A_{43},$$

and so, under this rotation, (A_{23}, A_{31}, A_{12}) transform like the components $M_1=A_{23}$, $M_2=A_{31}$, $M_3=A_{12}$, of a vector **M** and (A_{41}, A_{42}, A_{43}) like the components (N_1, N_2, N_3) of another vector **N**. Since any rotation can be assembled from rotations about the axes this result is completely general. Any antisymmetric tensor F_{jk} can be expressed in terms of two 3-vectors **M** and **N** as

$$\begin{matrix} 0 & M_3 & -M_2 & N_1 \\ -M_3 & 0 & M_1 & N_2 \\ M_2 & -M_1 & 0 & N_3 \\ -N_1 & -N_2 & -N_3 & 0 \end{matrix} \quad = \quad F_{jk}. \qquad (7.30)$$

The converse is <u>not</u> true. The components of F must transform correctly under all Lorentz transformations, and those involving motion mix the components of **M** and **N** which must therefore be related vectors. An important example occurs in electromagnetism where the components of the electric and magnetic fields **E** and **B** form a covariant tensor F with $\mathbf{M} = \mathbf{B}$ and $\mathbf{N} = i\mathbf{E}/c$.

7.6 *Infinitesimal Lorentz Transformations*

The matrix

$$\Omega_{jk} = \begin{matrix} \cos\Omega & -\sin\Omega & 0 & 0 \\ \sin\Omega & \cos\Omega & 0 & 0 \\ 0 & 0 & 1 & 0 \\ 0 & 0 & 0 & 1 \end{matrix} \qquad (7.31)$$

describes a rotation through an angle Ω about the 3-axis and the matrix for a "boost" along the X_1 axis with a velocity $u = \beta c$, and therefore with $\gamma = (1 - \beta^2)^{-1/2}$, is

$$B_{jk} = \begin{matrix} \gamma & 0 & 0 & i\beta\gamma \\ 0 & 1 & 0 & 0 \\ 0 & 0 & 1 & 0 \\ -i\beta\gamma & 0 & 0 & \gamma \, . \end{matrix} \qquad (7.32)$$

If the angle Ω is small i.e. $d\Omega$, then, since $\cos d\Omega \to 1$ and $\sin d\Omega \to d\Omega$, the infinitesimal rotation matrix is $I_{jk}+d\Omega_{jk}$, where I_{jk} is the unit matrix $\mathrm{diag}(1,1,1,1)$, and the only non-zero elements of $d\Omega_{jk}$ are $d\Omega_{12}=-d\Omega$ and $d\Omega_{21}=d\Omega$. Similarly if $\beta \to d\beta$ then $\gamma(\beta) \to 1$ and the infinitesimal boost matrix is $I_{jk} + dB_{jk}$, where now the only non-zero elements of dB_{jk} are $dB_{14}=id\beta$ and $dB_{41}=-id\beta$. Both $d\Omega_{jk}$ and dB_{jk} are antisymmetric matrices and, in general, the matrix of any infinitesimal Lorentz transformation is the sum of the unit matrix and an infinitesimal antisymmetric matrix.

7.7 *Scalar Invariants*

Although relativistic equations written in 4-vector notation are usually quite concise and it is easy to see whether they are properly constructed from covariant quantities, solving specific relativistic problems is often much more tedious. This is particularly the case when the calculation begins and ends in the laboratory frame which frequently has no particular relation to the system under discussion, so that the calculation may involve a succession of Lorentz transformations. This type of calculation can sometimes be simplified by identifying scalar invariants, which will have the same value in all inertial reference frames.

If two particles of 4-momenta $p_j=(\mathbf{p},ie/c)$ and $P_j=(\mathbf{P},iE/c)$ collide, a quantity of major interest is the energy E'' in the centre of momentum frame, since this is the energy available to cause reactions. Consider then the invariant $Q^2 = (p_j+P_j)(p_j+P_j)$. In the centre of momentum frame F* we have $\mathbf{p}^*+\mathbf{P}^* = \mathbf{0}$ and so $p_j{}^*+P_j{}^* = (\mathbf{0},\{ie^*+iE^*\}/c) = (\mathbf{0},iE''/c)$. Thus the available energy $E'' = c(-Q^2)^{1/2}$, which can be expressed as $E'' = [(e+E)^2-c^2(\mathbf{p}+\mathbf{P}).(\mathbf{p}+\mathbf{P})]^{1/2}$.

For the second example we anticipate some results of Chapters 8 and 9 where we show that two scalar invariants that can be constructed from the electromagnetic fields in vacuum, are **E.B** and $\frac{1}{2}(\varepsilon_0 E^2 - B^2/\mu_0)$. These reveal that if **B**=0 in any frame then **B** is either zero or perpendicular to **E** in any other frame, and there is no other frame in which **B**≠0 but **E**=0. These two invariants are part of a wider set of results that exploit the similarity between a second rank tensor and a 4×4 matrix (see Appendix A5).

7.8 Discussion

In 4-vector notation we can see from the structure of an equation whether it is compatible with relativity. A manifestly covariant equation describes a relativistically invariant relation. Some classical concepts fit easily into this scheme, for example both momentum **p** and current density **J** (see #9.1) become the spatial parts of 4-vectors whose fourth components iE/c and $ic\rho$ have a familiar meaning.

We can also see that some relations such as $\nabla.\mathbf{J}+\partial\rho/\partial t=0$ then have an even more compact form $\partial J_j/\partial x_j=0$, that the relation $E^2=p^2c^2+m_0^2c^4$ is neatly expressed as $p_jp_j+m_0^2c^4=0$ and that the d'Alembertian operator $\Box^2=\nabla^2-c^{-2}\partial^2/\partial t^2$ becomes the scalar operator $\partial/\partial x_j\,\partial/\partial x_j$. Other old friends however turn out to be less manageable, in particular neither velocity nor force turns out to be part of a 4-vector. The simplest covariant quantities to which they are related are the 4-velocity $v_j=\gamma(v)(\mathbf{v},ic)=p_j/m_0$ and the Minkowski 4-force $F=\gamma(v)(\mathbf{F},iW/c)$ where $W=\mathbf{F}.\mathbf{v}$. Sadly this 4-force is not defined except in terms of the velocity of the particle on which it acts and this reduces its utility. It will clearly not be very useful in dealing with a group of particles with different velocities. Though these 4-vectors (in association with the proper time) can usefully describe the dynamics of a single particle, they are a poor substitute for the simple concepts of classical dynamics. Perhaps the most tedious aspect of 4-vector notation is that it deprives us of the vector product, so that when this product appears in a classical equation the relativistic generalisation involves 2nd rank tensors.

Some problems will have neat solutions in 4-vector notation and the notation is especially useful in general discussions of, for example, conservation laws. Nevertheless its prime function is to make manifest the invariance, or otherwise, of an equation claiming to express a universally valid physical relation. Once we have verified that a theory is properly covariant we may frequently prefer to choose a particularly convenient reference frame and revert to 3-vector notation. This is certainly the case in most applications of electromagnetism.

Problems

7.1 Find the matrix for an infinitesimal rotation $d\phi$ about the X_3 axis followed by an infinitesimal rotation through $d\theta$ about the X_1 axis.

7.2 Crystallographers often use non-orthogonal axes defined by vectors **a**, **b** and **c** of unequal length. The components of a vector **F** are defined by $\mathbf{F} = F_a\mathbf{a} + F_b\mathbf{b} + F_c\mathbf{c}$. Express **F.G** in terms of the components of **F** and **G**.

7.3 If each point x_β in an elastic medium is given a displacement u, the strain in the medium is described by $\sigma_{\alpha\beta} = ½\{[\partial u_\alpha/\partial x_\beta + \partial u_\beta/\partial x_\alpha]$. Show that this is a symmetric 2nd rank tensor. Relate the change in the density at x_β to the components of $\sigma_{\alpha\beta}$.

7.4 F_{ij} is a covariant 2nd rank 4-tensor, show that its trace F_{ii} (sum over i implied) is a scalar.

7.5 Show that the odd order invariants of an antisymmetric tensor are zero. See Appendix 5.

7.6 If F is the tensor in equation (7.30) constructed from the two 3-vectors **M** and **N**. Show that one of its even order invariants is $M^2 + N^2$ and that the other is $(\mathbf{M.N})^2$.

CHAPTER 8

CLASSICAL ELECTROMAGNETISM

Electromagnetism as Maxwell finally formulated it in 1864 was in fact, though this could hardly have been recognised at the time, an entirely relativistic theory. Some 40 years would elapse before Lorentz derived the coordinate transformation that left Maxwell's equations invariant, and before Einstein formulated the principle of relativity. The close relation between these developments and Maxwell's equations is well worth exploring.

We must remember that the founders of electromagnetism in the nineteenth century were unaware of the existence of a fundamental unit of electric charge and so had no qualms about treating electricity as a continuous fluid. The laws of electromagnetism that they developed form a closed structure, all of whose parts are linked together and cannot easily be treated piecemeal. Thus, although we follow tradition and begin with the laws of electrostatics and the inverse square law, we should remember that these laws, like Newton's laws of motion, are idealisations, and difficult to verify by direct experiments. Our trust in their validity is due to how precisely their direct consequences agree with practical experience and experiment (see appendix A7).

The basic electrostatic laws are that a set of bodies in a state of relative rest can exert attractive or repulsive forces on each other, that action equals reaction, that the forces act along the line of centres between the bodies and that the force varies inversely with the square of the distance between the bodies. A further important law is that each body can be assigned a **charge** q such that the forces acting on the body and exerted by the body are both proportional to q. In addition, in an isolated system, the algebraic sum of all the charges remains constant, in other words charge is conserved. Finally, the forces due to several bodies are linearly additive so that the force acting on a body at $\mathbf{r}$ with charge q, due to other bodies at $\mathbf{r}_n$ with charges q_n, is

$$\mathbf{F}(\mathbf{r}) = q\ \Sigma_n q_n(\mathbf{r} - \mathbf{r}_n)/4\pi\varepsilon_o|\mathbf{r} - \mathbf{r}_n|^3. \qquad (8.1)$$

The remaining equations of electrostatics, summarised by $\mathbf{F} = q\mathbf{E}$, $\mathbf{E} = -\nabla\phi$, $\nabla.\mathbf{E} = \rho/\varepsilon_o$, can be derived from Eq.(8.1) and, significantly for the development of physical theories, the whole of the laws of electromagnetism, as it applies to charges in vacuum, can plausibly be deduced from Eq.(8.1), combined with charge conservation and the Lorentz transformation. We will show this in the next chapter. An excellent extended account is given by Rosser (1968).

Historically, however, ideas about magnetism preceded both relativity and electrostatics by many centuries. The idea of a magnetic field with field lines goes back to the thirteenth century. It is therefore sensible to regard classical electromagnetism as a complete theory, with an existence entirely independent of special relativity.

A charged body at rest experiences a force $\mathbf{F}$ that can be expressed as $q\mathbf{E}$, the product of its charge q and an electric field $\mathbf{E}$. This serves to define $\mathbf{E}$ in mechanical terms. Charges in motion constitute currents and we can either define I the actual rate of charge transport or $\mathbf{J}$ the vector current density in Ampères/m². Magnetic field $\mathbf{B}$ can be defined either in terms of the force $d\mathbf{F} = Id\ell\times\mathbf{B}$, as a vector product acting at right angles to a current element $Id\ell$ in a wire, or in terms of a force density $\mathbf{f} = \mathbf{J}\times\mathbf{B}$ N/m³ in a region where the current density is $\mathbf{J}$, or by the Lorentz equation giving the force $q\mathbf{v}\times\mathbf{B}$ acting on a particle of charge q moving with a constant velocity $\mathbf{v}$:-

$$d\mathbf{F} = Id\ell\times\mathbf{B}, \quad (8.2a) \qquad \mathbf{f} = \mathbf{J}\times\mathbf{B}, \quad (8.2b) \qquad \mathbf{F} = q\mathbf{v}\times\mathbf{B}. \quad (8.2c)$$

The Biot-Savart and Ampère laws describe how currents, *and currents alone*, generate a magnetic field, and these laws are equivalent to $\nabla.\mathbf{B} = 0$, $\nabla\times\mathbf{B} = \mu_o\mathbf{J}$. The second equation is, however, not compatible with conservation of charge and the existence of electrostatic phenomena since, if $\mu_o\mathbf{J} \equiv \nabla\times\mathbf{B}$ then $\nabla.\mathbf{J} \equiv 0$, but charge conservation is expressed by $\nabla.\mathbf{J} + \partial\rho/\partial t = 0$ making $\partial\rho/\partial t$ identically zero so that variation in the charge anywhere in space would be impossible. Maxwell's resolution of this problem was to add $\varepsilon_o\dot{\mathbf{E}}$ to $\mathbf{J}$ in Ampère's law. This, in conjunction with $\nabla.\mathbf{E} = \rho/\varepsilon_o$, preserves charge conservation. This new term $\varepsilon_o\dot{\mathbf{E}}$, the displacement current (density), describes how changing electric fields are related to magnetic fields. With Faraday's law of induction, which describes how changing magnetic fields are related to electric fields, it is responsible for the fact that the electromagnetic equations have wave solutions of velocity $c=1/\sqrt{\mu_0\varepsilon_0}$.

The complete set of Maxwell's equations is

$$\nabla.\mathbf{B} = 0, \quad (8.3a) \qquad \nabla\times\mathbf{E} + \partial\mathbf{B}/\partial t = 0, \quad (8.3b)$$

$$\nabla.\mathbf{E} = \rho/\varepsilon_0, \quad (8.3c) \qquad \nabla\times\mathbf{B} = \mu_0\mathbf{J} + \mu_0\varepsilon_0\partial\mathbf{E}/\partial t. \quad (8.3d)$$

These together with the formula for the force density, or the Lorentz force acting in vacuum on a body with a uniform velocity $\mathbf{v}$ and charge q,

$$\mathbf{f} = \rho\mathbf{E} + \mathbf{J}\times\mathbf{B}, \quad (8.4a) \qquad \mathbf{F} = q\mathbf{E} + q\mathbf{v}\times\mathbf{B}, \quad (8.4b)$$

are the basic equations of classical electromagnetism. They can be supplemented with the auxiliary vectors $\mathbf{P}$, $\mathbf{M}$, $\mathbf{D}$ and $\mathbf{H}$ with $\mathbf{D} = \varepsilon_0\mathbf{E}+\mathbf{P}$ and $\mathbf{B} = \mu_0(\mathbf{H}+\mathbf{M})$ where $\mathbf{P}$ and $\mathbf{M}$ summarise the properties of continuous media in a convenient form. We note that conservation of charge is implicit in Eqs.(8.3c) and (8.3d).

Because $\nabla.\mathbf{B} = 0$ the magnetic field $\mathbf{B}$ can be expressed as the curl of a vector potential $\mathbf{A}$

$$\mathbf{B} = \nabla\times\mathbf{A}, \qquad (8.5a)$$

and, since Eq.(8.3b) gives $\nabla\times(\mathbf{E} + \partial\mathbf{A}/\partial t) = 0$, $\mathbf{E}$ can be expressed in terms of $\mathbf{A}$ and a scalar potential ϕ as

$$\mathbf{E} = -\partial\mathbf{A}/\partial t - \nabla\phi\,. \qquad (8.5b)$$

Thus $\mathbf{E}$ and $\mathbf{B}$ with 6 components can be expressed in terms of $\mathbf{A}$ and ϕ with only 4 components. These, when substituted in Eq.(8.3d), give $\nabla\times\nabla\times\mathbf{A} = \mu_0\mathbf{J} - \mu_0\varepsilon_0[\partial^2\mathbf{A}/\partial t^2 - \partial/\partial t(\nabla\phi)]$ and this, written in terms of Cartesian components A_λ etc. with λ=1,2 or 3, becomes

$$\nabla^2 A_\lambda - \mu_0\varepsilon_0\ddot{A}_\lambda = -\mu_0 J_\lambda + \nabla_\lambda(\nabla.\mathbf{A} + \mu_0\varepsilon_0\dot{\phi})\,. \qquad (8.6a)$$

Similarly Eq.(8.3c) becomes, after some rearrangement,

$$\nabla^2\phi - \mu_0\varepsilon_0\,\ddot{\phi} = -\rho/\varepsilon_0 - \partial/\partial t(\nabla.\mathbf{A} + \mu_0\varepsilon_0\dot{\phi})\,. \qquad (8.6b)$$

Adding the gradient $\nabla\psi$ of an arbitrary scalar ψ to $\mathbf{A}$ does not alter $\mathbf{B}$ but subtracts $\nabla(\partial\psi/\partial t)$ from $\mathbf{E}$. This can then be corrected by subtracting $\partial\psi/\partial t$ from ϕ. The effect on the two identical terms in brackets in (8.6a,b) is to subtract $\nabla^2\psi-\mu_0\varepsilon_0\partial^2\psi/\partial t^2$ and, since ψ is arbitrary, this can be used to cancel the unwanted term $\nabla.\mathbf{A} + \mu_0\varepsilon_0\dot{\phi}$. Thus we can find a vector potential $\mathbf{A}$ and a scalar potential ϕ giving fields satisfying Maxwell's equations with prescribed source terms $\mathbf{J}$ and ρ, and also satisfying both the gauge condition

$$\nabla.\mathbf{A} + c^{-2}\partial\phi/\partial t \equiv \nabla.\mathbf{A} + \mu_0\varepsilon_0\partial\phi/\partial t = 0, \qquad (8.7a)$$

and the two inhomogeneous wave equations

$$\nabla^2 A_\lambda - c^{-2}\partial^2 A_\lambda/\partial t^2 = -\mu_0 J_\lambda, \tag{8.7b}$$

$$\nabla^2 \phi - c^{-2}\partial^2 \phi/\partial t^2 = -\rho/\varepsilon_0. \tag{8.7c}$$

These wave equations have, as explicit solutions, the retarded potentials:-

$$\mathbf{A}(\mathbf{R},T) = \mu_0 \iiint \mathbf{J}(\mathbf{r},t)\,dV(\mathbf{r})/(4\pi|\mathbf{R}-\mathbf{r}|), \tag{8.8a}$$

$$\phi(\mathbf{R},T) = \iiint \rho(\mathbf{r},t)\,dV(\mathbf{r})/(4\pi\varepsilon_0|\mathbf{R}-\mathbf{r}|), \tag{8.8b}$$

where, in the integral, $t = T-|\mathbf{R}-\mathbf{r}|/c$ and T is the retarded time; so that a time $|\mathbf{R}-\mathbf{r}|/c$ elapses before a change in the charge distribution at $\mathbf{r}$ is felt at $\mathbf{R}$ (see Appendix A7).

The field equations yield Poynting's identity

$$\nabla.(\mathbf{E}\times\mathbf{B})/\mu_0 + \mathbf{E}.\mathbf{J} + \varepsilon_0\mathbf{E}.\dot{\mathbf{E}} + \mathbf{B}.\dot{\mathbf{B}}/\mu_0 = 0, \tag{8.9}$$

in which $\mathbf{E}.\mathbf{J}$ is the rate (per unit volume) at which the field does work on charged bodies, the terms in $\varepsilon_0\mathbf{E}.\dot{\mathbf{E}}$ and $\mathbf{B}.\dot{\mathbf{B}}/\mu_0$ are the rates of growth of the electric and magnetic field energy densities, $\mathbf{N} = \mathbf{E}\times\mathbf{B}/\mu_0$ is Poynting's energy flux vector, and Eq.(8.9) is the energy conservation equation.

In vacuum the field transfers momentum to charged particles (in unit volume) at a rate $\mathbf{f} = \rho\mathbf{E} + \mathbf{J}\times\mathbf{B}$ and the field equations can be used to eliminate the charge and current densities and express the force density in terms of the fields alone. The result is

$$\mathbf{f} + \partial/\partial t(\varepsilon_0\mathbf{E}\times\mathbf{B}) = \varepsilon_0[\mathbf{E}(\nabla.\mathbf{E})-\mathbf{E}\times(\nabla\times\mathbf{E})]+[\mathbf{B}(\nabla.\mathbf{B})-\mathbf{B}\times(\nabla\times\mathbf{B})]/\mu_0.$$

On the left $\varepsilon_0\mathbf{E}\times\mathbf{B}$ is $\mathbf{N}/c^2$, where $\mathbf{N}$ is Poynting's vector, and thus (see #6.1) must be the field momentum density. Using Cartesian components E_ν, B_λ, etc. we can write the whole equation as

$$f_\nu + (1/c^2)\partial N_\nu/\partial t = \partial T^M_{\nu\lambda}/\partial r_\lambda, \tag{8.10}$$

where, on the right hand side, the second rank three-dimensional tensor $T^M_{\nu\lambda}$ (Maxwell's stress tensor) is

$$T^M_{\nu\lambda} = \tfrac{1}{2}[\varepsilon_0(E_\nu E_\lambda+E_\lambda E_\nu-E^2\delta_{\nu\lambda})+(1/\mu_0)(B_\nu B_\lambda+B_\lambda B_\nu-B^2\mu_0\delta_{\nu\lambda})]. \tag{8.11}$$

It gives the ν th component of the force acting across a surface element dS_λ and since, on the left hand side of Eq.(8.10), we have the sum of the rates of increase of matter and field momentum density, Eq.(8.10) expresses the momentum conservation law just as (8.9) expresses the energy conservation law. Because momentum is a vector, whereas energy is a scalar, it is more complicated. We will see later how relativity combines the two laws in one equation.

According to Eq.(8.8a) the vector potential at $\mathbf{r}$ due to a very slowly moving charge q at the origin, with a velocity $v \ll c$ is, apart from a delay r/c,

$$\mathbf{A}(\mathbf{r}) = \mu_o q\mathbf{v}/4\pi r \tag{8.12a}$$

and the scalar potential is

$$\phi(\mathbf{r}) = q/4\pi\varepsilon_o r. \tag{8.12b}$$

If the charge has an acceleration $\mathbf{a}$ the electric field at $\mathbf{r}$ is

$$\mathbf{E}(\mathbf{r}) = -\dot{\mathbf{A}} - \nabla\phi = -\mu_o q\mathbf{a}/4\pi r - q\mathbf{r}/4\pi\varepsilon_o r^3, \tag{8.12c}$$

and the dominant term at large distances is the acceleration dependent first term which falls off as $1/r$ whereas the second term falls off as $1/r^2$.

From (8.12c) the energy flux in the direction $\mathbf{r}$ at $\mathbf{r}$ is

$$N_r = (\varepsilon_o/\mu_o)^{1/2}E_\perp^2 = (\varepsilon_o/\mu_o)^{1/2}[E^2-(\mathbf{E}.\mathbf{r})^2/r^2] = (\mu_o/\varepsilon_o)^{1/2}q^2[a^2r^2-(\mathbf{a}.\mathbf{r})^2]/(4\pi r^2c)^2 \tag{8.13}$$

and the total radiated power is

$$W = (\mu_o/\varepsilon_o)^{1/2}q^2a^2/(6\pi c^2). \tag{8.14}$$

We end our brief survey of classical electromagnetic theory by noting that whereas the vacuum fields **E** and **B** fit easily into relativity, this will not be the case for the polarisation and magnetisation vectors **P** and **M** within a material medium, for these vectors are material properties. They are associated with a division of charge, in material media, into mobile charge and bound charge. The latter, though it may be displaced within an atom or molecule cannot, unlike mobile charge, be displaced a macroscopic distance without destroying the structure of the medium. The vectors **P** and **M**, together with $\mathbf{D} = \varepsilon_o\mathbf{E}+\mathbf{P}$ and $\mathbf{H} = \mathbf{B}/\mu_o-\mathbf{M}$, since they may be associated with the motion of material objects must therefore be handled with great care in relativistic calculations.

We also note that Eqs.(8.9) and (8.10) indicate that the electromagnetic field can store energy and momentum. They also show how these stored quantities are related to the fields, which themselves have independent definitions.

Perhaps the clearest indication of the intrinsically relativistic nature of classical electromagnetism is furnished by the formulae (8a) and (8b) for the retarded potentials. They show that if charge is displaced at $\mathbf{r}$ the effect is only felt at $\mathbf{R}$ after a time $t=|\mathbf{R}-\mathbf{r}|/c$ (which is independent of any motion) has elapsed.

Problems

8.1 Show that in a region where ρ and **J** are zero, Maxwell's equations have plane transverse wave-solutions of velocity c.

8.2 Show that a wave with the single electric field component $E_y = E_o \cos \pi x/2a \exp[i(\omega t - \beta z)]$ where E_o, a, ω and β are constant is a solution of Maxwell's equations. Obtain the relation between ω, β and a, and show that the phase velocity ω/β is greater than c and the group velocity $d\omega/d\beta$ is less than c. Find the components of the magnetic field **B** associated with this wave, and the time average value of Poynting's vector $\langle \mathbf{N} \rangle$ on the z axis. Find the ratio of $\langle \mathbf{N} \rangle$ to the energy density on the axis.

8.3 Confirm that the retarded potentials (8.8a,b) satisfy the inhomogeneous wave equations (8.7b,c).

8.4 The inner conductor of a vacuum spaced coaxial line carries a current I and is at a potential V relative to the outer conductor. Show, by integrating Poynting's vector over the annular region between the two conductors, that a power IV flows in the direction of the current.

8.5 A straight, enamelled, copper wire carries a steady current I and is immersed in a fluid medium of relative permeability μ in which there is a uniform <u>applied</u> field $\mathbf{B}_o$ perpendicular to the axis of the wire. Both the copper and the enamel have unity relative permeability. Show that the <u>impressed</u> magnetic field within the wire will be $\mathbf{B} = 2\mu\mathbf{B}_o/(1+\mu)$, but nevertheless the force per unit length acting on the wire is IB. (You will need to evaluate the Maxwell stress tensor at the surface of the enamel.)

8.6 Show that the r.m.s. electric field due to thermal radiation at room temperature is several hundred Vm^{-1}.

CHAPTER 9

ELECTROMAGNETISM and RELATIVITY

In this chapter we intend to exhibit the new insights into the structure and applications of electromagnetism that can be obtained by considering the theory in a relativistic context. Although we already know that Maxwell's equations must be invariant under Lorentz transformations, we must first confirm this by expressing all the electrodynamic equations in a manifestly covariant form.

9.1 Covariance of the Basic Equations

In Appendix 1 we show that $dx_1dx_2dx_3dx_4 \equiv dx_1dx_2dx_3 icdt$ is an invariant scalar. If ρ is a static charge density the charge $dq = \rho dx_1dx_2dx_3$ in a volume element, must remain the same under a transformation to a new frame in which the charge is in motion. Thus dq must be a scalar invariant and ρ must therefore transform like $x_4 = ict$. Consider then the effect on a 4-vector $J_k = (0,0,0,ic\rho)$ of a transformation to a new frame moving with a velocity $-u$ along the x_1 axis. In this new frame the charge moves with a velocity $+u$ and, classically, constitutes a current density ρu. The Lorentz transformation gives $J_k' = \gamma(u)(\rho u,0,0,ic\rho)$. The Lorentz contraction has increased the charge density to $\rho' = \gamma\rho$ so that $J_k' = (\rho' u,0,0,ic\rho')$. $J_1' = \rho' u$ is clearly the current density in the new frame and so the 3-dimensional current density $\mathbf{j}$ and ρ can be combined to form a covariant 4-vector

$$J_i = (\mathbf{j}\ , ic\rho), \tag{9.1a}$$

and then the invariant equation

$$\Sigma_i \partial J_i/\partial x_i = 0 \tag{9.1b}$$

expresses charge conservation.

In Chapter 8 we have seen that the vector and scalar potentials satisfy the inhomogeneous wave equations

$$\nabla^2 A_\kappa - c^{-2}\partial^2 A_\kappa/\partial t^2 = -\mu_0 j_\kappa \tag{9.2a}$$

$$\nabla^2 \phi - c^{-2}\partial^2 \phi/\partial t^2 = -\rho/\varepsilon_0 \tag{9.2b}$$

where we now recognise $\nabla^2 - c^{-2}\partial^2/\partial t^2 \equiv \Box^2$ as an invariant scalar operator. If we put

$$A_i = (\ \mathbf{A},\ i\phi/c) \tag{9.3}$$

then Eqs.(9.2a,b) can together be written as

$$\Box^2 A_i = -\mu_o J_i . \tag{9.4}$$

Since the operator is a scalar and J_i is a 4-vector, we can deduce that A_i must also be a 4-vector. We note further that the gauge condition

$$\nabla.\mathbf{A} + (\partial\phi/\partial t)/c^2 = 0$$

can also be written in the covariant form

$$\partial A_i/\partial x_i = 0. \tag{9.5}$$

Because A_i is a 4-vector, quantities such as $\partial A_i/\partial x_j$ are the components of a 2nd rank tensor and so

$$F_{ij} = \partial A_j/\partial x_i - \partial A_i/\partial x_j \tag{9.6}$$

is an antisymmetric tensor. The diagonal elements are, of course, zero, but consider $F_{23} = -F_{32} = \partial A_3/\partial x_2 - \partial A_2/\partial x_3 = B_1$. Similarly $F_{12} = -F_{21} = B_3$ and $F_{31} = -F_{13} = B_2$. The component $F_{41} = -F_{14} = \partial A_1/\partial x_4 - \partial A_4/\partial x_1 = (1/ic)(\partial A_1/\partial t + \partial\phi/\partial x_1) = iE_1/c$, and similarly $F_{42} = -F_{24} = iE_2/c$ and $F_{43} = -F_{34} = iE_3/c$, so that F_{ij} can be expressed entirely in terms of the fields **E** and **B**. It is known as the *field tensor*, and has the form

$$F_{ij} = \begin{matrix} 0 & B_3 & -B_2 & -iE_1/c \\ -B_3 & 0 & B_1 & -iE_2/c \\ B_2 & -B_1 & 0 & -iE_3/c \\ iE_1/c & iE_2/c & iE_3/c & 0 \end{matrix} . \tag{9.7}$$

The factors i appear because we chose to define x_4 as ict so that we could write the fundamental invariant as $x_j x_j$, and the factors c appear because of our (S.I.) choice of units. We leave it to the reader to confirm that

$$\partial F_{ij}/\partial x_j = \mu_o J_i, \tag{9.8}$$

and that this contains the two <u>inhomogeneous</u> Maxwell equations with the 4 components $\nabla\times\mathbf{B} - \mu_o\varepsilon_o\partial\mathbf{E}/\partial t = \mu_o\mathbf{j}$ and $\nabla.\mathbf{E} = \rho/\varepsilon_o$.

As long as we regard **E** and **B** as expressed in terms of **A** and ϕ the two homogeneous Maxwell equations are identities, but because **E** and **B** have an operational definition, whereas **A** and ϕ are to some extent arbitrary and their physical meaning less intuitively obvious, it is important and useful to find a covariant formulation of the two <u>homogeneous</u> equations $\nabla\times\mathbf{E} + \partial\mathbf{B}/\partial t = 0$ and $\nabla.\mathbf{B} = 0$. The result, as we shall see, is rather clumsy, and this in part explains why using the potentials is often much more convenient.

The field tensor is antisymmetric and so, if j=k, the expression $\partial F_{ij}/\partial x_k + \partial F_{ki}/\partial x_j$ is identically zero. If i=j the first term is zero and the second term can be written as $-\partial F_{jk}/\partial x_i$, thus the expression $\partial F_{ij}/\partial x_k + \partial F_{ki}/\partial x_j + \partial F_{jk}/\partial x_i$

is zero if any two subscripts are equal. If however, i=1, j=2, k=3 it is

$$\partial F_{12}/\partial x_3+\partial F_{31}/\partial x_2+\partial F_{23}/\partial x_1 = \partial B_3/\partial x_3+\partial B_2/\partial x_2+\partial B_1/\partial x_1 = \mathbf{\nabla}.\mathbf{B},$$

and, if i=1, j=2, k=4 it is

$$(1/ic)[\partial B_3/\partial t - (\partial E_1/\partial x_2 - \partial E_2/\partial x_1)] = (1/ic)[\mathbf{\nabla}\times \mathbf{E} + \partial \mathbf{B}/\partial t]_3.$$

Similar combinations give the other components of $\mathbf{\nabla}\times\mathbf{E}+\partial\mathbf{B}/\partial t$. The invariant 3rd rank tensor equation

$$\partial F_{ij}/\partial x_k + \partial F_{ki}/\partial x_j + \partial F_{jk}/\partial x_i = 0 \qquad (9.9)$$

is therefore a covariant statement of the two homogeneous field equations. With this we have completed putting Maxwell's field equations into a manifestly covariant form, though not, it must be admitted, one that looks likely to be of much use in practical applications.

9.2 The Force Equation

The equation for the force density is $\mathbf{f} = \rho\mathbf{E} + \mathbf{j}\times\mathbf{B}$, and has components such as $f_1 = \rho E_1 + j_2B_3 - j_3B_2$. Since the field tensor F_{ij} is antisymmetric with no diagonal components F_{ii}, we can write the first three (spatial) components of the right hand side as $F_{\kappa j}J_j$ where κ = 1, 2, or 3 but j is summed from 1 to 4. Now the force density **f**, together with w the rate of work per unit volume, form the 4-vector $f_i = (\mathbf{f}, iw/c)$ and since $w = \mathbf{E}.\mathbf{j} = -icF_{4j}J_j$, we see that the force density equation can be combined with the power density equation in a single covariant 4-vector equation

$$f_i = F_{ij}J_j, \qquad (9.10)$$

where now i =1,2,3 or 4 and j is summed from 1 to 4.

Since all the results of electrodynamics can be obtained from the field equations and the force density equation, we have confirmed that the whole theory and the principles of special relativity are consistent, and that the equations of electrodynamics can be expressed in a manifestly covariant form. However we do not have to use this form in working relativistic problems involving only the fields and charged particles pursuing prescribed trajectories unless it is convenient to do so. The 3-dimensional form of Maxwell's field equations is covariant, even though not manifestly so. The situation is quite different when we want to discuss the motion of charged particles in response to the field, for classical dynamics is not relativistically invariant.

Although force density can be expressed as a 4-vector, the Lorentz force $\mathbf{F} = q\mathbf{E} + q\mathbf{v}\times\mathbf{B}$ needs more modification. We have to use the proper time $\tau = t/\gamma(v)$ associated with the moving body, its 4-velocity v_i, its 4-momentum p_i and the 4-force $F_i = dp_i/d\tau$ to obtain a covariant equation

$$F_i \equiv dp_i/d\tau = qF_{ij}v_j. \qquad (9.11)$$

The reader should check that, for $v \ll c$, the first three components reduce to $\mathbf{F} = q\mathbf{E} + q\mathbf{v}\times\mathbf{B}$ and that the fourth component is $de/dt = q\mathbf{E}.\mathbf{v}$, where e is the energy of the particle.

9.3 The Transformation of the Current Density

In the course of a practical calculation it will often be convenient to do most of the calculation in one frame, perhaps the rest frame of a charged particle, and only at the end convert the results to the laboratory frame. We will therefore need to know how **E** and **B** transform. This is not difficult, since the fields appear as components of the antisymmetric covariant tensor F_{ij}, but we will consider first the covariant current density 4-vector J_i.

In a change of reference frame in which $x_i \rightarrow x_i' = a_{ij}x_j$ we must have $J_i' = a_{ij}J_j$ though, because the spatial part of J is a 3-vector, we need not investigate rotations but only the simple "boost", in which the primed coordinate system C' moves with a velocity $u = \beta c$ along the X axis of system C. We can ignore the unchanged transverse components J_2 and J_3, so that the only important elements of a_{ij} will be $a_{11}=\gamma(u)$, $a_{12}=i\beta\gamma$, $a_{41}=-i\beta\gamma$, $a_{44}=\gamma$. This leads to

$$j_1' = \gamma(j_1 + i\beta J_4) = \gamma(j_1 - u\rho), \qquad \rho' = \gamma(\rho - uj_1/c^2), \qquad (9.12)$$

where $\mathbf{j}$ is the ordinary 3-dimensional current density with components j_1, j_2, j_3. For $u \ll c$ the longitudinal component gives the obvious result $j_1' = j_1 - \rho u$. But, to apparently the same approximation, $\rho' = \rho - j_1 u/c^2$ is a less obvious result. However, suppose that in the coordinate system C the current is due to charge of density ρ moving with a velocity v along the x axis, then $j_1 = \rho v$, so that $\rho' = \gamma(u)(\rho - uv\rho/c^2) = \gamma(u)\rho(1-uv/c^2)$ and the term $1-uv/c^2$, as well as $\gamma(u)-1$, is quadratic in $1/c$. Thus the approximation $\rho' = \rho - uj_1/c^2$ is only apparently of first order. Note, however, the exact result $j_1' \equiv \rho' v'$, where $v' = (v-u)/(1-uv/c^2)$ is the velocity of the charge with respect to C'.

9.4 The Field Transformation

Since the fields are 3-vectors we can again ignore rotations and discuss only the effect of a boost transformation. We start by using

$$F_{ij}' = a_{ip}a_{jq}F_{pq} \tag{9.13}$$

and with $a_{11} = \gamma$, $a_{14} = i\beta\gamma$, $a_{22} = a_{33}=1$, $a_{41} = -i\beta\gamma$, $a_{44} = \gamma$ we have, for example, $B_3' \equiv F_{12}' = a_{1p}a_{2q}F_{pq} = a_{11}a_{22}F_{12}+a_{14}a_{22}F_{42} = \gamma F_{12} + i\beta\gamma F_{42} = \gamma[B_3-uE_2/c^2]$. Proceeding in this way we obtain

$$B_1'= B_1, \quad B_2'= \gamma(B_2+uE_3/c^2), \quad B_3'= \gamma(B_3-uE_2/c^2), \tag{9.14a}$$

$$E_1'= E_1, \quad E_2'= \gamma(E_2-uB_3), \quad E_3'= \gamma(E_3+uB_2). \tag{9.14b}$$

These results can also be expressed in vector notation in terms of the parallel and perpendicular components of the fields as

$$\mathbf{B}_{/\!/}' = \mathbf{B}_{/\!/}, \quad \mathbf{B}_{\perp}' = \gamma(u)(\mathbf{B}_{\perp} - \mathbf{u}\times\mathbf{E}/c^2), \tag{9.15a}$$

$$\mathbf{E}_{/\!/}' = \mathbf{E}_{/\!/}, \quad \mathbf{E}_{\perp}' = \gamma(u)(\mathbf{E}_{\perp} + \mathbf{u}\times\mathbf{B}). \tag{9.15b}$$

The field components parallel to the relative velocity are unchanged, while the transverse components are not only changed but also become mixed. This is hardly surprising, for if the field is due to charges at rest in F they are in motion in F' and represent currents which generate magnetic fields. Notice also that, as we expect, when u/c is small and $\gamma(u)\rightarrow 1$ the expressions for E_2' and E_3' resemble the formulae for the induced field in a moving conductor, i.e. $E_2' = E_2 - uB_3$ and $E_3'= E_3 + uB_2$.

In using these results we must remember that they refer to fields at a unique point in space-time which is specified by x_i in F and $x_i' = a_{ij}x_j$ in F'. If, for example, we have a point charge q at the origin in F' the field $\mathbf{B}'$ in F' is zero and, with $x_i' = (\mathbf{r}',ict')$, the electric field is simply $\mathbf{E}' = q\mathbf{r}'/4\pi\varepsilon_0 r'^3$. In F a typical field component is

$$E_1(r,t) = E_1' = qr_1'/4\pi\varepsilon_0[r_1'^2+r_2'^2+r_3'^2]^{3/2} = \gamma q(r_1-ut)/4\pi\varepsilon_0[\gamma^2(r_1-ut)^2+r_2^2+r_3^2]^{3/2}. \tag{9.16a}$$

On the line of motion, where $r_2=r_3=0$, the only non-zero component is

$$E_1(r_1,t) = q\ /[4\pi\varepsilon_0\gamma^2(u)(r_1-ut)^2]. \tag{9.16b}$$

Because of the factor $\gamma^2(u)$ in the denominator, E_1 on the line of motion becomes small as $u\rightarrow c$, except for the singularity at $r_1 = ut$, which describes the present position of the particle. On the other hand, off the line of motion, but in the plane containing the point charge, the transverse components of **E** and **B**, although they fall off as the inverse square of the distance, are proportional to $\gamma(u)$ and so become larger as $u \to c$.

It is perhaps worth remarking that these results had earlier been calculated directly from Maxwell's equations without invoking the Lorentz transformation, by Lienard (1898) and Wiechert (1900), using the retarded potentials. However the relativistic derivations, even though they are sometimes not immediately obvious, are rather simpler than the direct classical calculations.

As a trivial example of the field transformation we consider a conducting sphere with a uniform magnetisation **M** rotating slowly about its axis of magnetisation with an angular velocity ω.

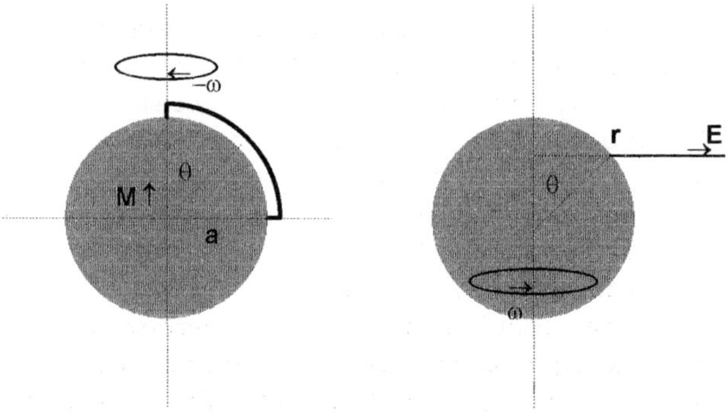

Figure 9.1a Figure 9.1b
The Unipolar Generator

Consider first the case when the sphere is at rest and, as shown in Figure 9.1a, a circular segment of wire makes contact with the sphere at the pole and at the equator, and rotates around the sphere with an angular velocity $-\omega$. The radial component of **B** at the angular position θ is $2/3\ \mu_0 M\cos\theta$ and the velocity of an element of wire, of length $a d\theta$, is $-a\omega\sin\theta$. Thus the emf induced in this element is

$dV = 2/3\,\omega a^2\mu_o M\sin\theta\cos\theta\,d\theta$ and the total induced e.m.f. is $V = 1/3\,\omega a^2\mu_o M$. Since the magnetic field within the sphere is $\mathbf{B} = 2/3\,\mu_o\mathbf{M}$ we can express the e.m.f. as $\frac{1}{2}\omega a^2 B$.

Next consider the wire as fixed and the sphere as rotating. The velocity of an element of the sphere at $\mathbf{r}$ is $\boldsymbol{\omega}\times\mathbf{r}$ and in its instantaneous rest frame the only field is $\mathbf{B}$. In the laboratory frame, Eq.(9.15b) gives (with $\mathbf{u}=-\boldsymbol{\omega}\times\mathbf{r}$), $\mathbf{E}_{/\!/} = 0$, $\mathbf{E} = \mathbf{E}_\perp = -\gamma(\omega r)(\boldsymbol{\omega}\times\mathbf{r})\times\mathbf{B}$ and so, ignoring terms of order $(\omega r/c)^3$, $\mathbf{E} = -(\boldsymbol{\omega}\times\mathbf{r})\times\mathbf{B}$. This gives $\nabla\times\mathbf{E} = \nabla\times[\mathbf{B}\times\boldsymbol{\omega}\times\mathbf{r})] = \nabla\times[\boldsymbol{\omega}(\mathbf{B}.\mathbf{r})] = \mathbf{B}\times\boldsymbol{\omega} = 0$ [where we have used $\nabla\times\mathbf{r} = 0$, the fact that $\boldsymbol{\omega}$ and $\mathbf{B}$ are both constant vectors and that $\boldsymbol{\omega}$ is parallel to $\mathbf{B}$]. We can therefore express $\mathbf{E}$ as $-\nabla\phi$. We see that its magnitude is ωBr, that it is perpendicular to both $\mathbf{v} = \boldsymbol{\omega}\times\mathbf{r}$ and to $\mathbf{B}$, and so has the direction shown in Fig. (9.1b). This can now be used to calculate the potential difference between a point on the equator and a point on the axis i.e. at the pole. The result is $\phi_e-\phi_p = \int\omega Br\,dr = \frac{1}{2}\omega a^2 B$, in agreement with the result for the e.m.f. in the segment of wire when the wire rotates about the fixed sphere. Thus the situation is in entire agreement with our intuitive notion of what relativity is all about. For an informative and much fuller discussion of the unipolar generator, the reader should consult Rosser (1968).

9.5 Field Invariants

An antisymmetric tensor, such as the field tensor F_{ij}, is associated with two scalar invariants, $F_{ij}F_{ij}$, and the determinant $|F|$ (see Appendix A5); these two invariants are $K_2 = 2[F_{12}{}^2 + F_{13}{}^2 + F_{14}{}^2 + F_{23}{}^2 + F_{24}{}^2 + F_{34}{}^2] = 4[B^2-E^2/c^2]$, and $K_4 = |F| = (F_{12}F_{34} + F_{23}F_{14} + F_{31}F_{24})^2 = -(\mathbf{E}.\mathbf{B}/c^2)^2$. It will be convenient to replace these by

$$I_1 = \tfrac{1}{2}(\varepsilon_o E^2-B^2/\mu_0) \tag{9.17a}$$

$$I_2 = \mathbf{E}.\mathbf{B}. \tag{9.17b}$$

The reader may wish to verify that these are invariants by using the field transformation Eqs.(9.15a,b). Eq.(9.17a) shows that if the electric field energy density locally exceeds the magnetic field energy density in any frame then it does so in every frame. Predominantly electric or predominantly magnetic fields retain these properties in all frames. Equation (9.17b) shows that unless $\mathbf{E.B} = 0$, either because one of the fields is zero or because the fields are at right angles, then neither **E** nor **B** can vanish in any frame. Equally if we start with a purely electrostatic field then, in any other frame, **E** is non-zero and **B** will be either zero or perpendicular to **E**. Thus these two formal results give a very useful understanding of the invariant properties of the electromagnetic field.

In most cases there will be a frame of reference in which **E** is parallel to **B**, so that Poynting's vector $\mathbf{E}\times\mathbf{B}/\mu_o = 0$, and there is no energy flux. To show this we choose the OX_1 axis and the direction of motion to be perpendicular to both **E** and **B**, so that $E_1' = E_1 = B_1' = B_1 = 0$. The only component of $\mathbf{N}' = \mathbf{E}'\times\mathbf{B}'$ in F' is then

$$N_1' = E_2'B_3' - E_3'B_2' = \gamma^2[(E_2 - uB_3)(B_3 - uE_2/c^2) - (E_3 + uB_2)(B_2 + uE_3/c^2)]. \tag{9.18a}$$

This will vanish if $u/(1+u^2/c^2) = (\mathbf{E}\times\mathbf{B})_1/(B^2+E^2/c^2)$ or, with a more general choice of axes, if

$$2\mathbf{u}/(1+u^2/c^2) = 2\mathbf{E}\times\mathbf{B}/(B^2+E^2/c^2) = (\mathbf{E}\times\mathbf{B}/\mu_o)/\tfrac{1}{2}(\varepsilon_o E^2 + B^2/\mu_o). \tag{9.18b}$$

Notice that the right hand side is the ratio of the energy flux to the energy density and must therefore be related to a group velocity. Let θ be the angle between **E** and **B**, so that $\mathbf{E}\times\mathbf{B} = EB\sin\theta$ and let $E = \zeta cB$, then we have

$$u/(1+u^2/c^2) = c\sin\theta/(\zeta + 1/\zeta) \tag{9.18c}$$

and this yields $u < c$ unless both $\theta = 90^o$ *and* $\zeta = 1$ when $u = c$. Thus unless both $\mathbf{E.B} = 0$ and $E^2 - B^2c^2 = 0$ we can always, using a velocity u less than c, find a frame in which **B** is parallel to **E** and the energy flux $\mathbf{E}\times\mathbf{B}/\mu_o$ vanishes.

The exceptional case corresponds to locally plane waves, for then $\mathbf{E} = \mathbf{E}_o f(\mathbf{n.r} - ct)$ and $\mathbf{B} = \mathbf{B}_o f(\mathbf{n.r} - ct)$ with $\mathbf{n}\times\mathbf{E} = c\mathbf{B}$, $\mathbf{n.B} = \mathbf{n.E} = 0$ and $\mathbf{E} = c\mathbf{B}\times\mathbf{n}$ (where **n** is the unit vector in the direction of propagation). We see that $\mathbf{E.B} = 0$ and $E^2 = c^2B^2$, so that for a plane wave both invariants are zero and there is no frame in which the energy flux $\mathbf{E}\times\mathbf{B}/\mu_o$ can be transformed away. Furthermore, when the two invariants are zero, $|\mathbf{E}\times\mathbf{B}|^2 = E^2B^2 - (\mathbf{E.B})^2 = E^2B^2 = (E^2 - c^2B^2)B^2 + c^2B^4 = c^2B^4$, and we see that $\mathbf{E}\times\mathbf{B}/\mu_o = \tfrac{1}{2}c(\varepsilon_o E^2 + B^2/\mu_o)$, thus the energy velocity, given by Eq. (9.18b), is c and invariant, which clearly agrees with the basic principle of special relativity.

Although a plane wave cannot be completely transformed away and although its velocity remains c, it does not mean that its energy flux is invariant. By considering a plane wave with $B_2 = E_3 = 0$, but $E_2 = cB_3$, we can use Eq.(9.18a) to calculate Poynting's vector N' in a frame travelling with a velocity u in the direction of the wave. The result is

$$N' = N(1-u/c)/(1+u/c), \tag{9.19}$$

Thus as $u \to c$ both the energy flux and the energy density tend to zero but their ratio remains c. If, for example, we are receding from a star with a velocity u and the star emits light of frequency f, not only is the frequency of each photon Doppler shifted to $f' = f(1-u/c)^{1/2}/(1+u/c)^{1/2}$ but the rate n' at which photons arrive is reduced in the same ratio so that, in conformity with (9.19) the energy flux is reduced to $n'hf' = nhf(1-u/c)/(1+u/c)$. This result combined with Hubble's law and the recession of distant galaxies, is believed to explain why the night sky is dark (Olber's paradox).

9.6 *Polarisation and Magnetisation*

A satisfactory and self-consistent treatment of the connection between the macroscopic electromagnetic vectors **P** and **M** and the microscopic or atomic structure of matter is not, as is often erroneously assumed, a matter of some ill-defined quantum or statistical-mechanical average over huge numbers of particles, for that would certainly eliminate the possibility of observing macroscopic fluctuations arising from the discrete quantal or statistical structure of matter, fluctuations which are in fact readily observable experimentally. Rather, the connection involves introducing the notion of a finite resolution associated with any macroscopic measurement (Robinson 1971, 1973). We need not however bother with the connection between the microscopic and macroscopic worlds, but instead ask what transformation properties the vectors **P** and **M** must have to make relativistic sense of expressions such as $\mathbf{D} = \varepsilon_0\mathbf{E} + \mathbf{P}$. It is immediately clear that **M** and **P** together, and **H** and **D** together, must form antisymmetric second-rank tensors of the same form as the field tensor F_{ij}. We will call these tensors K_{ij} and G_{ij}. The first row of K must be $0, M_3, -M_2, icP_1$

and the first row of G must be $0, H_3, -H_2, -icD_1$. We have the covariant relation

$$G_{ij} = (F_{ij}/\mu_o - K_{ij}), \tag{9.20}$$

and G satisfies the equation

$$\partial G_{ij}/\partial x_j = J_i, \tag{9.21}$$

which corresponds to the two Maxwell equations

$$\nabla\times\mathbf{H} - \dot{\mathbf{D}} = \mathbf{j}, \qquad \nabla.\mathbf{D} = \rho. \tag{9.22}$$

Neither G nor K satisfy anything corresponding to the homogeneous equations $\nabla.\mathbf{B} = 0$ and $\nabla \times \mathbf{E} + \partial\mathbf{B}/\partial t = 0$, for the vectors **P** and **M** involve quite arbitrary, and possibly quite complicated, material properties. In general they depend on the thermodynamic state of the material medium, and even on its recent thermal history. We will, however, confine our attention to the very simplest case in which **M** is zero and $\mathbf{P}=\chi\,\varepsilon_o\mathbf{E}$ and $\mathbf{D}=(1+\chi)\varepsilon_o\mathbf{E}$ or $\mathbf{D}=\varepsilon\varepsilon_o\mathbf{E}$. When we consult a table of physical constants to find the numerical value of ε, or the refractive index $n=\varepsilon^{1/2}$, we expect to find the value appropriate to the rest frame of the medium or, more strictly, its centre of mass frame. There may be a further complication since each of the components of **E** and **D** is part of a 2nd rank 4-tensor so that ε, which relates **D** and **E**, could be a tensor of rank 4, but we will assume that ε is a simple constant. We now look at refraction in a moving medium.

9.7 *Fizeau's Drag Coefficient*

In 1851 Fizeau investigated the velocity of light in water flowing with a velocity *u*; had he been dealing with sound with a velocity s in water at rest he would have expected to get $s+u$ or $s-u$. For light he found, from his experiments, that $c'=(c/n)\pm(1-1/n^2)u$. The factor $1-1/n^2$ is known as Fizeau's drag coefficient.

We will suppose that we have plane waves with all the electric vectors parallel to the 2-axis and all the magnetic vectors parallel to the 3-axis and, for brevity, we will omit all the component subscripts. In the rest frame of the water $E'=\gamma(u)(E-uB)$, $B'=\gamma(B-uE/c^2)$, and so $D'=\varepsilon\varepsilon_o\gamma(E-uB)$, $\mu_o H'=\gamma(B-uE/c^2)$. In the laboratory frame $D=\gamma(D'+uH'/c^2)$ and $H=\gamma(H'+uD')$, so that $D=\varepsilon_o\gamma^2[(n^2-\beta^2)E-(n^2-1)\beta cB]$ and $\mu_o H=\gamma^2[(1-n^2\beta)^2B+(n^2-1)\beta E/c]$, where $\beta=u/c$. These equations yield D and B in terms of E and H as

$$D = [n^2\varepsilon_0(1-\beta^2)E - (n^2-1)\beta H/c]/[1-n^2\beta^2], \qquad (9.23a)$$
$$B = [\mu_0(1-\beta^2)H - (n^2-1)\beta E/c]/[1-n^2\beta^2]. \qquad (9.23b)$$

For a plane wave, $\nabla\times\mathbf{E} + \partial\mathbf{B}/\partial t = 0$ and $\nabla\times\mathbf{H} - \partial\mathbf{D}/\partial t = 0$ give, in terms of frequency ν and wave-length λ, the relations $\nu B = E/\lambda$ and $\nu D = H/\lambda$ and when these results are combined with Eqs.(9.23a,b) we obtain a secular equation which yields the wave velocity as

$$c' = \nu\lambda = [(n^2-1)u \pm nc(1-\beta^2)]/[n^2-\beta^2] \qquad (9.24a)$$

or, when $u \ll c$,

$$c' = (n^2-1)u/n^2 \pm c/n = (1-1/n^2)u \pm c/n, \qquad (9.24b)$$

where the $\pm$ sign now corresponds to propagation with or against the flow rather than, as in Fizeau's formula, flow with or against the direction of propagation. Otherwise Eq.(9.24b) is in complete agreement with Fizeau's result.

If, instead of using a drag coefficient, we use an effective refractive index n^* the result is $n^* = n \pm u(n^2-1)/c$, in the laboratory frame.

9.8 *Radiation by a Charged Particle*

The classical Larmor result for the total power radiated by a particle of charge q with an acceleration a and a velocity much less than c is kq^2a^2/c^2, where, in S.I. units, $k = (\mu_0/\varepsilon_0)^{1/2}/6\pi$. This is essentially the power radiated in the instantaneous rest frame. The energy radiated in dt is

$$dU = kq^2a^2dt/c^2 = k(q^2/m^2c^2)(dp/dt)^2dt \qquad (9.25)$$

and, as this result (expressed in terms of the momentum p and the rest mass m) is exact in the instantaneous rest frame, it forms a convenient point of departure in our search for a covariant formula.

The momentum 4-vector is $p_i = (\mathbf{p}, ie/c)$ and thus we have $(dp_j/dt)(dp_j/dt) = (dp/dt)^2 + (dp_4/dt)^2 = (dp/dt)^2 - (de/dt)^2/c^2$. Since $de/dt = \mathbf{v}.d\mathbf{p}/dt$ this is approximately $(dp/dt)^2(1-v^2/c^2)$ and so, in the rest frame, we can replace $(dp/dt)^2$ by $(dp_j/dt)(dp_j/dt)$ and also replace t by the proper time τ. Then dU consists of a scalar quantity, multiplying the final dt. This is satisfactory since idU/c is the fourth component of the covariant radiated momentum $dP_i = (d\mathbf{P}, idU/c)$ and $icdt$ is the fourth component dx_4 of x_j. Therefore Eq.(9.25) must be the fourth component of the covariant 4-vector equation

$$dP_i = k(q/mc^2)^2(dp_j/d\tau)(dp_j/d\tau)dx_i. \qquad (9.26)$$

In the rest frame the spatial components of (9.26) are correctly zero, for in this frame the Larmor formula gives a symmetric (dipolar) polar diagram which cannot allow the radiation to carry off any momentum. Because (9.26) is covariant we are no longer restricted to the rest frame and we can use (9.26) to discuss the radiation from a particle moving with a velocity **v** relative to the laboratory. In the laboratory frame $dt=\gamma d\tau$ and $dp_i=\gamma m(\mathbf{v},ic)=\gamma mc(\boldsymbol{\beta},i)$ where $\boldsymbol{\beta}=\mathbf{v}/c$. Thus, with $Wdt=dU=-icdP_4$, the 4th component of (9.26) yields

$$W = k(q/c)^2\gamma^2(v)[(d\gamma\mathbf{v}/dt).(d\gamma\mathbf{v}/dt)-c^2(d\gamma/dt)^2], \qquad (9.27a)$$

This can be expressed, in terms of $\boldsymbol{\beta}=\mathbf{v}/c$, as

$$W = kq^2\gamma^6(\beta)[1-\beta^2(d\beta/dt)^2+(\boldsymbol{\beta}.d\boldsymbol{\beta}/dt)^2], \qquad (9.27b)$$

where $k=(\mu_o/\varepsilon_o)^{1/2}/6\pi$, or, in terms of the classical Larmor power W_L and the angle θ between the acceleration and the velocity, as

$$W = \gamma^6 W_L(1-\beta^2\sin^2\theta). \qquad (9.27c)$$

This gives the energy radiated by the particle per unit of laboratory time and it agrees exactly with the result derived directly from Maxwell's equations nearly a hundred years ago by Lienard (1898) and Wiechert (1900) (see Heitler 1935).

For a given velocity and <u>acceleration</u> the radiated power is γ^2 greater when the acceleration and the velocity are parallel than when they are perpendicular. But if v and $\gamma(v)$ are large, the acceleration in the laboratory frame can only be small and will tend to zero as $v\to c$. For a given <u>force</u> the radiation is greater by a factor $\gamma^2(v)$ when **a** and **v** are perpendicular to each other.

The spatial components of P_i give the radiated momentum, and since dx_i in Eq.(9.27) is the particle coordinate in the laboratory frame we can replace its space components by $\mathbf{v}dt$ and this gives

$$d\mathbf{P}/dt = W\mathbf{v}/c^2. \qquad (9.28)$$

As $v\to 0$ the radiated momentum tends to zero and the polar diagram becomes symmetric, with the intensity varying as $\sin^2\theta$, where θ is measured from the direction of the acceleration. As $v\to c$, $P\to W/c$, which is the appropriate form for a narrow beam of radiation in the forward direction of motion. There are important applications of these results to particle accelerators, free electron lasers and the generation of synchrotron radiation.

To illustrate their magnitude we consider a particle in a circular orbit of radius r in a fixed magnetic field B. Since the acceleration is v^2/r the power radiated can be expressed as $k(q\gamma^2 v^2/cr)^2$ which gives the energy radiated per revolution as $\Delta U = (\mu_o/\varepsilon_o)^{1/2}(q\gamma^2 v/c)^2 v/3r$ and, as $v \to c$, this becomes $\Delta U = (q^2/3\varepsilon_o r)(E/mc^2)^4$, where E is the particle energy. With ΔU expressed in MeV and E in Gev this is about $\Delta U = E^4/9r$ for electrons, thus electrons of 3 Gev energy in an orbit of radius 9m would lose about 1Mev per revolution. This is a significant limitation on the use of cyclic machines to accelerate electrons. Protons whose mass is 1840 times bigger would have neglible energy loss under the same conditions. For a fuller treatment of synchrotron radiation see Jackson (1975) and, for its relation to the free electron laser, Luchini and Motz (1990).

9.9 *The Thomas Precession*

The origin of the intrinsic spin $s = \hbar/2$ of the electron, and its anomalous moment $\mu = g(se/2m)$ with $g \approx 2$, were eventually explained by Dirac as a direct consequence of relativistic quantum mechanics, but the idea had been introduced some years earlier, in a rather *ad hoc* way, by Uhlenbeck and Goudsmit to explain the multiplet structure of spectral lines. The value of the moment was chosen to fit the observed Zeeman splittings of the lines in a magnetic field; however, with this moment, the calculated fine structure intervals were twice the observed value. The explanation was given by Thomas (1926).

If the electron has an orbital velocity $\mathbf{u}$ then the electric field $\mathbf{E}$ due to the nucleus leads to a magnetic field in the rest frame of the electron $\mathbf{B}' = -\gamma\mathbf{u}\times\mathbf{E}/c^2$. In this field the electron spin obeys the equation of motion

$$(d\mathbf{s}/d\tau)' = \mu\times\mathbf{B}' = -(ge/2mc^2)\mathbf{s}\times(\mathbf{u}\times\gamma\,\mathbf{E})$$

and so

$$(d\mathbf{s}/dt)' = -(ge/2mc^2)\mathbf{s}\times(\mathbf{u}\times\mathbf{E}). \qquad (9.29)$$

But (see sections 2.6 and 3.10) the rest frame axes are rotating with respect to a fixed inertial frame at a rate

$$\boldsymbol{\omega} = \gamma^2\mathbf{a}\times\mathbf{u}/c^2(\gamma+1) \qquad (9.30)$$

where $\mathbf{a} = e\mathbf{E}/m\gamma$ is the electron's acceleration. Thus in the fixed frame the equation of motion is

$$d\mathbf{s}/dt=(d\mathbf{s}/dt)'+\omega\times\mathbf{s} = (e/mc^2)\,\mathbf{s}\times[\mathbf{E}\times\mathbf{u}]\,[\tfrac{1}{2}g-\gamma/(\gamma+1)]. \qquad (9.31a)$$

Atomic electrons have values of γ near unity and thus, with g=2, the equation of motion becomes

$$d\mathbf{s}/dt \sim (e/2mc^2)\,\mathbf{s}\times(\mathbf{E}\times\mathbf{u}). \qquad (9.31b)$$

The r.h.s. has just half the value of the corresponding term in Eq.(9.29). This agrees with the experimentally observed fine structure splittings.

9.10 *g*-2 Experiments and a Test of Special Relativity

Dirac's relativistic quantum theory leads to an electron spin of $\tfrac{1}{2}\hbar$ and a magnetic moment $e\hbar/m_o c$ corresponding to a gyromagnetic ratio 2. Quantum electrodynamic corrections increase this to $g=2(1+a)$ where a = 0.001 159 657.... The excellent agreement between the experimental and the theoretical values of a is one of the reasons for the oft repeated statement that quantum electrodynamics is the "best theory we have" (see e.g. Aitchison and Hey 1982). Here however we follow Newman, Ford, Rich and Sweetman(1978) and use the combination of the results of a g-2 experiment made with very low energy electrons by Van Dyck, Schwinberg and Dehmelt (1977), and an experiment made by Wesley and Rich (1971) with 110keV electrons ($\gamma=1.22, v/c=0.57$), to test the basic dynamic relation between momentum p, rest mass m_o and velocity v.

In a magnetic field B the (orbital) cyclotron frequency is

$$\omega_c = eB/\Gamma m_o c \qquad (9.32a)$$

where

$$\Gamma = (p/m_o)/(dE/dp). \qquad (9.32b)$$

The spin precession frequency is

$$\omega_s = geB/2m_o c + (1-\gamma)\,\omega_c, \qquad (9.32c)$$

in which the first term is due to the direct interaction of the spin magnetic moment $\mu = ge\mathbf{s}/2m_o c$ with the magnetic field and the second comes from the Thomas precession, a purely kinematic effect, so that here $\gamma \equiv 1/(1-u^2/c^2)^{1/2}$. To get this term from Eq.(9.30), put $a=\omega_c u$ and $u^2/c^2= 1-1/\gamma^2$.

The quantity measured in the experiments is

$$\omega_d = \omega_s - \omega_c = (g/2 - \gamma/\Gamma)\,eB/m_o c. \qquad (9.32d)$$

The low energy experiments gave a=0.001 159 652 41 ±(20) and the experiments at 110keV gave a=0.001 159 657 70 ±(350)

so that taken together they gave $1 - \gamma/\Gamma = (5.3 \pm 3.5)\times 10^{-9}$ thus verifying the basic relation of relativistic dynamics to a few parts in a billion at $u = 0.57c$.

9.11 Electrostatics, Relativity and Electromagnetism

It is sometimes claimed that magnetism is no more than a relativistic correction to electrostatics and we now investigate what truth, if any, there is in this statement.

The electrostatic laws can be summarised (see Chapter 8) as

$$\nabla.\mathbf{E} = \rho/\varepsilon_o, \quad \mathbf{f} = \rho\mathbf{E}, \text{ and } \nabla\times\mathbf{E} = 0, \tag{9.33a}$$

which allow us to write $\mathbf{E} = -\nabla\phi$ and

$$\nabla^2\phi = -\rho/\varepsilon_o. \tag{9.33b}$$

These equations as they stand are clearly not covariant, but the charge conservation equation $\nabla.\mathbf{j} + \partial\rho/\partial t = 0$ can immediately be expressed covariantly as $\partial J_i/\partial x_i = 0$ in terms of the 4-vector $J_i = (\mathbf{j}, ic\rho)$. Poisson's equation (9.33b) is, however, only meant to apply to a <u>static</u> situation and in that context it would be unchanged if ∇^2 were replaced by $\Box^2 = \nabla^2 - (1/c^2)\partial^2/\partial t^2$. Eq. (9.33b) could then be written, using this invariant scalar operator, as

$$\Box^2 i\phi/c = -ic\mu_o\rho. \tag{9.34a}$$

Since $ic\rho$ is the fourth component of J_i, we can treat $i\phi/c$ as the fourth component of a 4-vector A_i which will have the components $(A_1, A_2, A_3, i\phi/c)$. We thus replace Poisson's equation by $\Box^2 A_4 = -\mu_o J_4$. We have, as yet, no idea about any possible significance that the spatial components of A_i may have, but we do know that J_i satisfies $\partial J_i/\partial x_i = 0$, and so we would require $\partial/\partial x_i\{\Box^2 A_i\} = 0$, and this is certainly satisfied if

$$\partial A_i/\partial x_i = 0. \tag{9.34b}$$

This relation, which, with hindsight, we may recognise as the Lorentz gauge condition, is not the only possibility, but it unquestionably makes clear the close connection between charge conservation, as expressed by $\partial J_i/\partial x_i = 0$, and the existence of the Lorentz gauge.

We also see that $E_i = ic\partial A_4/\partial x_i$. Furthermore, in static fields, $ic\partial A_i/\partial x_4 \equiv \partial A_i/\partial t = 0$, and we could, if we wished, express the electric field as

$$E_i = ic(\partial A_4/\partial x_i \pm \partial A_i/\partial x_4). \tag{9.34c}$$

With the minus sign, this conveniently vanishes when i=4 corresponding to the undefined quantity E_4. Altogether this suggests that we should perhaps consider the antisymmetric tensor

$$F_{ij} = \partial A_j/\partial x_i - \partial A_i/\partial x_j, \qquad (9.34d)$$

some of whose components (F_{i4} and F_{4i}) correspond to components of **E**. We note in passing that the the components of F_{ij} are unchanged if A_i is replaced by $A_i' = A_i + \partial Y/\partial x_i$, where Y is any scalar function, and also that the condition Eq.(9.34b), related to charge conservation, only requires that Y satisfy the homogeneous, and Lorentz invariant, wave equation $\Box^2 Y = 0$. Any "gauge" functions used to transform the potential which are also going to conserve charge, will have to satisfy this "gauge" equation.

We now have the outline of a possible covariant form of the static field laws and so we must next consider the force density equation $\mathbf{f} = \rho\boldsymbol{E}$. We know that **f** is the spatial part of a 4-vector (see Appendix A2.), and that ρ is the time component of the 4-vector J_i. If $\mathbf{f} = \rho\boldsymbol{E}$ is to be part of a covariant equation then $\boldsymbol{E}$ can only be the mixed space-time part of a 2nd rank tensor. Indeed we find that when $\mathbf{j} = 0$ the electrostatic force law can be written in terms of the anti-symmetric tensor F_{ik} as $f_i = F_{ik}J_k$, where k is summed from 1 to 4. It will not matter whether i takes on the values 1 to 3, or 1 to 4, since $F_{44} = 0$ and the only component of J_i is J_4. Furthermore, even if **j** is not zero, the fourth component of f_i comes out correctly as $iw/c = i\mathbf{E}.\mathbf{j}/c$. If $\mathbf{j}\neq 0$ there are additional spatial components of f_i and, because F_{ik} is antisymmetric, these can be written in terms of a new vector with components $B_1 = F_{23}$, $B_2 = F_{31}$, and $B_3 = F_{12}$ as $\mathbf{j}\times\mathbf{B}$. These forces, which depend on the flow of the charged fluid, are just what we would predict from our knowledge of the laws of relativistic dynamics, for if we start in a frame of reference in which all the charge is at rest and there are only electrostatic forces, a change to a moving frame is bound to lead to velocity dependent forces.

It should now be becoming clear that when we have tied a few loose ends together, we will arrive at all the usual equations of electrodynamics. We must realise however that we have certainly not achieved this by an inevitable succession of logical steps. At several points, for example in choosing the antisymmetric form $F_{ik} = \partial A_k/\partial x_i - \partial A_i/\partial x_k$, rather than the symmetric form $\partial A_k \partial x_i + \partial A_i/\partial x_k$, for the field tensor, we were undoubtedly guided by our prior knowledge of our ultimate goal. This is made obvious if we consider gravitation, which starts like electrostatics with a simple inverse square law but certainly does not lead to anything analogous to magnetism. In any case once we have obtained the electromagnetic equations by this formal process we still have to recognise the new velocity dependent effects as magnetism, i.e. the practical subject first described in terms of magnetic poles and lines of force by Maricourt in 1269. What we have done is little more than reverse the steps that confirmed that classical electromagnetism is a Lorentz invariant, relativistic theory.

It would appear from this approach that magnetic effects are only a relativistic correction to electrostatics. This theoretical conceit might be true if physics were only concerned with assemblies of particles of like charge e.g. an electron beam, but it leaves us with the difficulty of explaining why magnetic forces are so vastly more important than electrostatic forces in engineering. Power stations use machinery made from large masses of copper and ferromagnetic iron, rather than electrostatic generators such as the Van der Graaf machine. Here we abruptly recollect the peculiar importance of the existence in the real world of positively and negatively charged particles.

Let us begin by considering the scalar invariant $E^2 - c^2B^2$. It seems that starting with an electrostatic field we could never produce a pure magnetic field, yet it is a commonplace that pure magnetic fields exist. Our difficulty arises from a misunderstanding; magnetostatic fields do *not* arise from the uniform motion of a set of charged particles even if some of them have different signs, rather they arise because particles with charges of different signs can have different *velocities* and there is *no* rest frame *common* to both types of particle.

In copper wire, the positively charged ion cores are, at least on average, at rest in the laboratory frame whereas the negative electrons are on average in motion, even if only very slowly. The electrostatic effects of the two types of charge cancel very exactly but, because of the vast numbers of mobile electrons (about 5.10^{28} per m^3 in a typical metal, giving a mobile charge density of about 10^9 coulomb m^{-3}), the magnetic effects can be very large. In a large electromagnet the electrostatic stress is entirely negligible but the magnetic stress can easily exceed 1000 atmospheres. It is doubtful if anyone knowing only the electrostatic laws and relativity would have anticipated the nature of a modern power station. On the other hand the structures of atoms, and of liquids and solids, are almost entirely determined by electostatic forces, with magnetism playing only a very minor role.

9.12 *Discussion*

We end our treatment of relativistic electromagnetism somewhat arbitrarily at this point, not because no other topics remain for discussion but rather because there are so many. We have not considered Cerenkov radiation from a charged particle moving through a refracting medium with a velocity exceeding the velocity of light in the medium; nor have we treated the important topic of the radiation produced in collisions between energetic particles; nor yet the reaction of the radiation field back on an accelerated particle; nor the effects of a particle's self-field on its dynamics. Topics such as these or a detailed theory of the free electron laser deserve a more extended treatment than could be given here and, in any case, there are excellent books dealing with these subjects, by Jackson (1975), Rohrlich (1965) and Luchini and Motz (1990).

We have also, although we have mentioned it several times as perhaps the most important consequence of special relativity, omitted to explain how imposing relativistic invariance on quantum mechanics leads to the existence of electron spin and thus, through the Dirac equation, determines the laws of chemistry and the structure of the natural world. A proper treatment of this topic would unfortunately require a fuller discussion of quantum mechanics than would be appropriate in a text on relativity. Nevertheless, because of its importance we give a brief sketch at the end of Chapter 11.

Problems

9.1 Show that a transformation to a frame moving with a velocity u in the direction of a plane wave alters both E and B in the ratio $[(1-u/c)/(1+u/c)]^{1/2}$.

9.2 In its rest frame a body has no magnetisation M and only a single component of polarisation P_2 . Show that in a frame in which it has a velocity u along the x_1 axis it will have a component M_3 of magnetisation.

9.3 A plane circularly polarised wave propagating along the x_3 axis has field components which are the real parts of $E_1 = E_0 e^{i(\kappa x_3 - \omega t)}$, $E_2 = iE_1$, $B_1 = -ic^{-1}E_2$ and $B_2 = c^{-1}E_1$ where E_0 is a constant amplitude. Show that the time average of the angular momentum in any region symmetric about the x_3 axis is zero.

9.4 An approximately plane wave of large but finite transverse dimensions is described by replacing E_0 in problem 9.3 by a slowly decreasing function $E_0(x_1,x_2)$ of the transverse coordinates x_1 and x_2. Since $\mathbf{E}$ must satisfy $\nabla.\mathbf{E}=0$ there will be a component $E_3=(i/k)(\partial E_0/\partial x_1+i\partial E_0/\partial x_2)e^{i(\kappa x_3-\omega t)}$, but $\mathbf{B}$ will still be given by $\mathbf{B} = -i\mathbf{E}/c$. Find an expression for the time average of the angular momentum per unit length along the x_3 axis and compare it with the time average energy per unit length.

9.5 What is the maximum change in the refractive index due to Fizeau drag that might be observed in water flowing at 7m/s (the value used by Fizeau)? How long a path in the water would be needed to give an appreciable fringe shift due to motion?

9.6 Suppose that all the mobile electrons (about 2.10^{29} with a total charge of 3.10^{10}C) were removed from a metal sphere of radius 1m. What would be the electric field at a distance of 1km from the sphere?

9.7 A copper wire of radius 10mm carries a current of 1000A. Estimate the drift velocity of the conduction electrons and the magnetic field at the surface of the wire.

9.8 Beams of non-relativistic electrons in vacuum diverge because of electrostatic repulsion, show that this effect vanishes as their velocity approaches c.

CHAPTER 10

RELATIVISTIC DYNAMICS III

The usual classical force **F** does not form part of a 4-vector without modification but we can, as long as we are only discussing the force acting on a single particle, use the proper time associated with the particle and express the equation of motion as $dp_i/d\tau = F_i$ in terms of the Minkowski 4-force $F_i = \gamma(v)(\mathbf{F}, i\mathbf{F}.\mathbf{v}/c)$. However, since this introduces the velocity **v** of the particle on which the force acts, it would obviously be an impossible procedure for dealing with a system of interacting particles, each with its own velocity and proper time. The first sections of this chapter are devoted to devising a technique for dealing with this problem based on the fact that, unlike the force, the force density, i.e. the force acting per unit volume on matter, is the spatial component of a 4-vector.

10.1 Continuous Media

Although, at least on an atomic scale, any real material medium has a grain and a discrete structure we can often treat it as continuous. Macroscopic physics deals with calculations and measurements of finite resolution and if this resolution is coarser than the grain of the medium we do not detect the grain directly but only those properties of the medium that are coarser than its grain. These properties are not averages and indeed they may exhibit macroscopic fluctuations arising, as for instance in shot noise, from the underlying structure (in this case the discrete charge of the electron), but they are smoothed macroscopic functions of the time and the spatial coordinates. If Θ is some macroscopic parameter or property we can consider its space and time derivatives $d\Theta/dx$ and $d\Theta/dt$ and assume that macroscopically, dx and dt can be treated as infinitesimal long before they are as small as the grain of the medium. Alternatively, if the actual microscopic description of the system is expressed in terms of Fourier expansions, information about the microscopic graininess of the medium is encoded by Fourier components whose spatial or temporal frequencies are too high to be detected in any macroscopic measurement of limited resolution. A simple example may perhaps clarify this idea. On close inspection wet sand is seen to consist of small grains of solid matter mixed with water, but if we happen to tread on quick-sand we are chiefly conscious of it as a viscous, continuous, and not very buoyant, medium.

On the other hand a field such as the electromagnetic field is a truly continuous system characterised by, for example, its energy and momentum densities. We now intend to extend this idea to material systems.

10.2 The Energy-Momentum Tensor

If **f** is the force density i.e. the force acting per unit volume, then the rate of change of the momentum of the matter in a volume dxdydz will be $d\mathbf{p}/dt = \mathbf{f}dxdydz$, and because $icdt = dx_4$ we could write this as $f_\lambda dx_1 dx_2 dx_3 dx_4 = icdp_\lambda$ where $\lambda = 1,2$ or 3. Since $dx_1 dx_2 dx_3 dx_4$ (see Appendix A1) is an invariant scalar the *three* components f_λ must transform like the three spatial components dp_λ of the covariant energy-momentum vector $dp_i = (d\mathbf{p}, idE/c)$, so that **f** is clearly part of a covariant 4-vector whose fourth component f_4, determined by $f_4 dx_1 dx_2 dx_3 = (i/c)dE/dt$ is, apart from the factor i/c, the rate w at which the force does work in unit volume. Thus we have obtained, as the covariant generalisation of force density **f**, a simple 4-vector

$$f_i = (\mathbf{f}, iw/c). \tag{10.1a}$$

The basic dynamical law is

$$f_i dx_1 dx_2 dx_3 dx_4 = icdp_i, \tag{10.1b}$$

and we must now try to relate dp_i to a density.

Consider a fluid, with a classical velocity **v** and momentum density **q** and a classical energy density e, in the presence of a classical force density **f**. The basic equations obeyed by the fluid are the momentum and energy balance equations

$$\partial(q_\lambda v_\mu)/\partial r_\mu + \partial q_\lambda/\partial t = f_\lambda \quad \text{and} \quad \partial(ev_\mu)/\partial r_\mu + \partial e/\partial t = \mathbf{f}.\mathbf{v} = f_\mu v_\mu,$$

where, as usual, λ and μ run only over the values 1 to 3.

Our aim is to combine these equations in a single covariant equation of the form $\partial K_{ij}/\partial x_j = f_i$, where i and j run from 1 to 4 and K_{ij} is a covariant 2nd rank tensor.

We begin with a single particle of rest mass m, velocity $\mathbf{v}$, momentum $\mathbf{p}$ and energy $E = mc^2\gamma(v)$. This particle carries the momentum component p_j in the k direction at a rate $p_j v_k$ and transports energy in the same direction at a rate Ev_k. Let us define $v_i = (\mathbf{v}, ic)$ which is not a covariant vector, and form the quantity $p_i v_j$ which is not a covariant tensor. However, apart from factors i and c, the 9 spatial components give the momentum flow, the three components $\mathbf{p}v_4$ are the momentum, the component $p_4 v_4$ is the energy and the three components $p_4\mathbf{v}$ are the energy flow.

Now consider many non-interacting particles with momenta $p^n{}_i$ and velocities $v^n{}_j$, and consider those particles in a small volume dV. In another frame F' the element d$V \to$ dV' and it contains the same particles with momenta $p^n{}_j{}'$ and velocities $v^n{}_j{}'$. The transformed momenta p' will be covariantly related to the p but we need not assume that this applies to the v'. Let $p_i v_j = \Sigma_n p^n{}_i v^n{}_j$ be the sum over particles in dV, and $\Sigma_n p_i{}' v_j{}'$ be the sum in dV', and consider $K_{ij}=p_i v_j dx_4/dx_1 dx_2 dx_3 dx_4$. Since the denominator is a scalar, K_{ij} transforms like $p_i v_j dx_4$. But $v_j dx_4=(\mathbf{v},ic)icdt=ic(d\mathbf{x},dx_4)$ which is a covariant 4-vector even though v_i is not. Thus K_{ij} transforms like $p_i x_j$ and so must be a covariant 2nd rank 4-tensor, and is still a covariant tensor if we write it in terms of v_j and the momentum density $\pi_i = p_i/d^3x$ as

$$K_{ij} = \pi_i v_j \tag{10.2}$$

Since $\pi_i v_j = p_i v_j/d^3x$, the tensor K_{ij} is symmetric and

$$K_{ij} = K_{ji}. \tag{10.3}$$

If we introduce the energy flux vector $\mathbf{s} = e\mathbf{v} = -ic\pi_4\mathbf{v}$ we can write the tensor as

$$K_{ij} = \begin{matrix} \pi_1 v_1 & \pi_1 v_2 & \pi_1 v_3 & ic\pi_1 \\ \pi_2 v_1 & \pi_2 v_2 & \pi_2 v_3 & ic\pi_2 \\ \pi_3 v_1 & \pi_3 v_2 & \pi_3 v_3 & ic\pi_3 \\ is_1/c & is_2/c & is_3/c & -e \end{matrix} \tag{10.4}$$

and then the symmetry of K implies that

$$\mathbf{s} = c^2\boldsymbol{\pi}. \tag{10.5}$$

The first of the space components of

$$X_i = \partial K_{ij}/\partial x_j \tag{10.6}$$

is $X_1 = \partial(\pi_1 v_1)/\partial x_1 + \partial(\pi_1 v_2)/\partial x_2 + \partial(\pi_1 v_3)/\partial x_3 + \partial\pi_1/\partial t$ and the first three terms of this give the flux out of unit volume of the first component of momentum , while the last term is the rate of change of the density of this momentum component. Thus X_1 must be the first component of the force density. The result for X_4 is $X_4 = i(\nabla.\mathbf{s} + \partial e/\partial t)/c$, so that $-icX_4 = w$ is the rate at which the force does work in unit volume. Thus $X_i = f_i$ and

$$\partial K_{ij}/\partial x_j = f_i \tag{10.7}$$

is the covariant form of Newton's law in continuous media.

The force density f_i can be expressed as the negative gradient of a field stress tensor T_{ij} possibly, but not necessarily, the electromagnetic stress tensor considered in the last chapter, so that $f_i = -\partial T_{ij}/\partial x_j$ and, all told, the result

$$\partial Z_{ij}/\partial x_j \equiv \partial(K_{ij} + T_{ij})/\partial x_j = 0 \tag{10.8}$$

expresses the overall momentum-energy conservation law.

10.3 The Symmetry of the Complete Energy-Momentum Tensor

The energy-momentum tensor K_{ij} is clearly symmetric from the way that it is constructed, but we need to prove that the stress tensor T_{ij} is also symmetric, since conservation of angular momentum is directly linked to the total symmetry. We begin with the purely spatial components $T_{\lambda\mu}$, where λ and μ run only from 1 to 3, and we give an argument (Nye 1957) to show that this part of the stress tensor must be symmetric with $T_{\mu\lambda} = T_{\lambda\mu}$. Consider a small cube with sides of length ℓ parallel to the axes, and the torque acting on it due to the stresses T_{12} and T_{21}, which exert forces $F_1 = T_{12}\ell^2$ and $-F_1$ on the two faces normal to the x_2 axis and $F_2 = -T_{21}\ell^2$ and $-F_2$ on faces normal to the x_1 axis, thus giving a total torque $C_3 = (T_{12}-T_{21})\ell^3$ about the x_3 axis. Now if the mass (energy) density is finite the moment of inertia of the cube varies as ℓ^5 and so, in the limit $\ell \to 0$, the torque would cause an angular acceleration tending to infinity as $1/\ell^2$. Only if $T_{12}-T_{21}$ is everywhere zero can the tensor T describe a stress that can persist for even an infinitesimal time.

Next we must consider the fourth row and column of T. The spatial part of T must remain symmetric under a Lorentz transformation to a new frame moving along the x_1 axis with a velocity u, and described by a matrix with $a_{11}=\gamma(u)$, $a_{14}=iu\gamma/c$, $a_{22}=a_{33}=1$, $a_{41}=-iu\gamma/c$, $a_{44}=\gamma$. This will lead to

$T_{12}'=\gamma T_{12}+iu\gamma T_{42}/c$ and $T_{21}'=\gamma T_{21}+iu\gamma T_{24}/c$ and so, since we must (as we have just seen) have $T_{12}'=T_{21}'$, it follows that we must also have $T_{24} = T_{42}$, making the whole stress tensor symmetric. Since K_{ij} is already symmetric this ensures that the complete energy-momentum tensor Z_{ij} is also symmetric.

The first three elements of the fourth column of Z are $ic\pi_v$, where $\boldsymbol{\pi}$ is now the total momentum density (matter + field). The (44) element is iec, with e the total energy density, and the first three elements of the fourth row are $Z_{4v}= is_v/c$, where $\mathbf{s}$ is the energy flux vector. The relation

$$\mathbf{s} = c^2\boldsymbol{\pi} \qquad (10.9)$$

between the energy flux vector $\mathbf{s}$ and the momentum density $\boldsymbol{\pi}$ is therefore completely general. This may seem strange, for the classical limits in a fluid of density ρ and velocity $\mathbf{v}$ would be $\mathbf{s} = \frac{1}{2}\rho_0 v^2\mathbf{v}$ and $\boldsymbol{\pi} = \rho_0\mathbf{v}$. We must, however, remember that the relativistic energy flux $\mathbf{s}$ includes the transport of energy $\rho_0 c^2$ associated with the rest mass density ρ_0. The classical energy flux is $\mathbf{s}_c = \mathbf{s}-\rho_0 c^2\mathbf{v}$ and, since $\mathbf{s} = \boldsymbol{\pi}c^2= \gamma\rho_0\mathbf{v}c^2$, we get the expected limit $\mathbf{s}_c = c^2\rho_0\mathbf{v}(\gamma -1)\rightarrow \frac{1}{2}\rho_0 v^2\mathbf{v}$, as $(v/c)\rightarrow 0$.

The total energy-momentum tensor Z and the two parts from which it has been constructed, K for the particles and T for the fields are all symmetric, but we must beware of thinking that it will always be easy to partition the total tensor into two symmetric parts recognisable as particle and field terms. Thus in electromagnetic problems involving polarisable or magnetisable media the polarisation vector **P** and the magnetisation **M** are formally part of the field but, at the same time, describe the configurations of charged particles and currents within the medium. Similar problems will arise in dealing with deformable media in deciding how much of the elastic energy belongs to matter and how much to fields, see for example Møller (1971).

10.4 Angular Momentum I

Angular momentum and its conservation plays an important role in quantum mechanics and in the classification of the states of atoms, nuclei, nucleons and the elementary particles. The origin of the intrinsic spin angular momentum of the elementary particles and its behaviour is bound up with relativity and Lorentz invariance and so, despite its complexity, the subject is something that we must consider.

In classical mechanics the angular momentum of a set of particles with linear momenta $\mathbf{p}(n)$ and coordinates $\mathbf{r}(n)$ is $\mathbf{J} = \sum_n \mathbf{r}(n)\times\mathbf{p}(n)$ and, if they interact only through central forces, for which action equals reaction, $\mathbf{J}$ is a constant of the motion. If we define $J_{\lambda\mu} = r_\lambda p_\mu - r_\mu p_\lambda$ then J_{23}, J_{31} and J_{12} correspond to J_1, J_2 and J_3. An obvious covariant generalisation is to introduce the 4-momenta $p_i(n)$ and the 4-vector coordinates $x_i(n)=(\mathbf{r}(n), ict)$ with the same $x_4=ict$ for all the particles in the system, and then define

$$J_{ij} = \sum_n x_i(n)p_j(n)-x_j(n)p_i(n) \qquad (10.10)$$

which is clearly an antisymmetric, covariant, second rank tensor. Its spatial components correspond to the classical angular momentum.

If we restrict our attention to free particles where $dp_i(n)/dt=0$ we have $dJ_{ij}/dt=\sum_n(p_j(n)dx_i(n)/dt-p_i(n)dx_j(n)/dt)$ and, since $p_i(n) = m_0(n)\gamma[v(n)]dx_i/dt$, each term in the sum is zero so that

$$dJ_{ij}/dt = 0. \qquad (10.11)$$

This does not look like a covariant equation, yet obviously it is true for any system of free particles in any reference frame.

The tensor J was defined in terms of the particle positions and momenta at a single instant t in the frame F and we need to know whether this definition is also valid in another frame $\tilde{F}$. In this new frame $x_i(n)$ at t in F appears as $\tilde{x}_i{}^n$ at $\tilde{t}^n$ and these times for the different particles are no longer equal. Let $\tilde{v}_i{}^n = d\tilde{x}_i{}^n/d\tilde{t} = (\tilde{\mathbf{v}}^n, ic)$. This is constant, and at $\tilde{t}$ the nth particle in $\tilde{F}$ is at $\tilde{\mathbf{x}}^n(\tilde{t}) = \tilde{\mathbf{x}}^n(\tilde{t}^n)+ \tilde{\mathbf{v}}^n(\tilde{t}-\tilde{t}^n)$. Consider one term in the sum which contributes to $\tilde{J}$ at $\tilde{t}$. We omit the label n except on $\tilde{t}$ and obtain

$$\tilde{x}_i(\tilde{t})\tilde{p}_j-\tilde{x}_j(\tilde{t})\tilde{p}_i = \tilde{x}_i(\tilde{t}^n)\tilde{p}_j-\tilde{x}_j(\tilde{t}^n)\tilde{p}_i+(\tilde{v}_i\tilde{p}_j-\tilde{v}_j\tilde{p}_i)(\tilde{t}-\tilde{t}^n),$$

where we have used the fact that $\tilde{p}_i$ is constant. Since $\tilde{p}_i{}^n = m_0{}^n\gamma(v^n)\tilde{v}_i{}^n$, the coefficient of $(\tilde{t}-\tilde{t}^n)$ on the r.h.s. is zero and thus $\tilde{J}$ at a single time $\tilde{t}$ in $\tilde{F}$ is

$$\tilde{J}_{ij} = \tilde{x}_i{}^n(\tilde{t})\tilde{p}_j{}^n-\tilde{x}_j{}^n(\tilde{t})\tilde{p}_i{}^n,$$

which is a tensor whose spatial components are the components of the angular momentum at the single instant $\tilde{t}$.

A typical mixed space time component of J is

$J_{14} = \Sigma_n x_1^n p_4^n - x_4 p_1^n = \Sigma_n icx_1^n E^n/c^2 - ictp_1^n$.

But E^n/c^2 can be replaced by the inertial mass of the nth particle, $m^n = m_o^n\gamma(v^n)$, and then, in terms of the coordinates $X_\kappa = M^{-1}\Sigma_n x_\kappa^n m^n$ of the centre of mass [where κ=1,2 or 3 and $M = \Sigma_n m^n$] and the total momentum of the system $\mathbf{P} = \Sigma_n \mathbf{p}^n$, we get

$$J_{k4}/ic = MX_k - tP_k. \qquad (10.12)$$

Thus the equation $dJ_{ij}/dt = 0$ gives, since P_k is constant,

$$MdX_k/dt = P_k \qquad (10.13)$$

so that the motion of the centre of mass is uniform. We see how, this result and the conservation of angular momentum both become part of the same equation in relativity. We also note that the position of the centre of mass at t=0 is

$$X_k(0) = J_{k4}/icM. \qquad (10.14)$$

It is tempting to seek to generalise these results by, for example, defining $X_i = \Sigma x_i^n m^n/M$ with i=1 to 4, but this is $X_i = [\Sigma_n x_i^n p_4^n]/[\Sigma_n p_4^n]$, which is clearly not a covariant 4-vector.

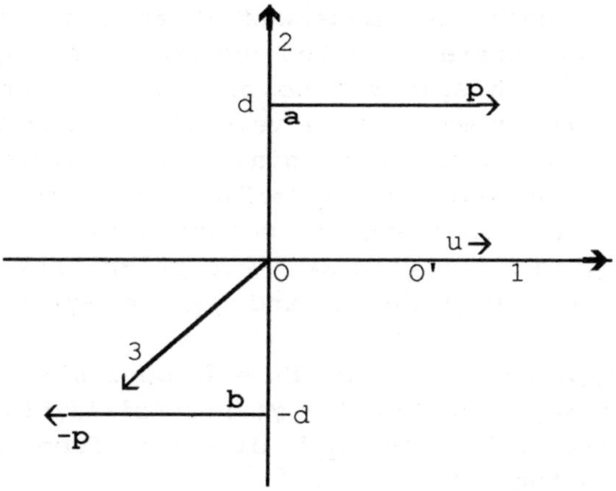

Figure 10.1 Centre of Mass and Angular Momentum

In practice, much of the complexity inherent in treating angular momentum relativistically arises from the difficulty of giving a covariant definition of the centre of mass coordinates. To understand the complex behaviour of X_k under a Lorentz transformation consider the two particles **a** and **b** in Figure 10.1, with equal and opposite momenta p and $-p$ parallel to the x_1 axis, and which, at t = 0, cross the x_2 axis at $x_2=d$ and $x_2=-d$. In this frame F the centre of mass is at rest at the origin and thus the only angular momentum component is $j_3 = -2pd$, that is $J_{12} = -J_{21} = -pd$. In the frame F', which moves along OX_1 with velocity u, the total mass is $M'=\gamma(u)M$ and the X_2' component of the centre of mass is $(J_{24}')/icM'$. But $J_{24}' = a_{22}a_{41}J_{21} = -iu\gamma J_{21}/c = iu\gamma J_{12}/c$, so that, at $t = t' = 0$,

$$X_2'(0) = uJ_{12}/Mc^2, \qquad (10.15)$$

and we see how the centre of mass has been moved off the $0X_1$ axis by an amount proportional to the angular momentum about the 3 axis in the original frame. The origin of this shift is obvious: in F' particle a has a velocity $(v-u)/(1-vu/c^2)$, which is less in magnitude than the negative velocity $-(v+u)/(1 + vu/c^2)$ of particle b; thus the mass of a is less than the mass of b and since they are at the same distance d on either side of the 0X axis the centre of mass is shifted towards b .

In F the centre of mass was at rest and all the angular momentum was internal to the system. We might describe it as a system with spin but no orbital angular momentum. In F' the centre of mass has a velocity $-u$ and it is displaced a distance X_2' from the origin. This gives the system an orbital angular momentum $L_{12}'=M'uX_2'=2\gamma px_2u^2/c^2=\gamma u^2J_{12}/c^2$.

The internal spin angular momentum is

$S_{12}' = -p_b'(d-X_2')+p_a'(d+X_2') = d(p_a'-p_b')+X_2'(p_a'+p_b')$ and, since $p_a' = \gamma(u)(p-uE/c^2)$ and $p_b' = -\gamma(u)(p+uE/c^2)$, this gives

$S_{12}' = 2\gamma pd-2\gamma(uE/c^2)(upd/E) = 2\gamma pd(1-u^2/c^2) = J_{12}/\gamma(u)$.

Altogether $J_{12}' = L_{12}'+S_{12}' = \gamma(u)[u^2/c^2+1/\gamma^2(u)]J_{12} = \gamma(u)J_{12}$. We can also calculate J_{12}' directly from $J_{12}' = a_{1i}a_{2j}J_{ij}$ which gives the same result.

This very simple example illustrates the complexity of relativistic angular momentum relations. The centre of mass is displaced by an amount which depends on the internal spin angular momentum in the initial frame and the initial separation of the total angular momentum into an internal spin part and an external orbital part associated with centre of mass motion is completely changed. We now try to incorporate field angular momentum within a covariant scheme.

10.5 Angular Momentum II

The fourth column of the total energy momentum tensor Z=T+K is $Z_{i4} = (ic\mathbf{q}, -e)$ where **q** and e are the momentum and energy densities. If therefore we define the third rank tensor

$$Y_{ijk} = x_i Z_{jk} - x_j Z_{ik} \tag{10.16}$$

components such as Y_{124} correspond to components of the angular momentum density **r**×**q**. This tensor satisfies

$$\partial Y_{ijk}/\partial x_k = Z_{ji} + x_i \partial Z_{jk}/\partial x_k - Z_{ij} - x_j \partial Z_{ik}/\partial x_k$$

and, since Z is symmetric and $\partial Z_{jk}/\partial x_k = \partial Z_{ik}/\partial x_k = 0$, we get

$$\partial Y_{ijk}/\partial x_k = 0, \tag{10.17}$$

and, if the integral is over the whole system,

$$J_{ij} = (1/ic)\iiint Y_{ij4} dV, \tag{10.18}$$

will be a covariant tensor. The space components are the components of the angular momentum and, because it is antisymmetric, there are no diagonal components. A typical space-time component is

$$J_{k4} = \iiint [i(e/c)x_k - ictq_k] dV = ic\iiint (\rho x_k - tq_k) dV$$

where e/c^2 has been replaced by the *inertial* mass density ρ. For an isolated system both the total mass $M = \iiint \rho dV$ and the total momentum $\mathbf{P} = \iiint \mathbf{q} dV$ are constant, and the centre of mass position is

$$X_k = (1/M)\iiint \rho x_k dV \tag{10.19}$$

so that

$$J_{k4} = ic[MX_k - tP_k]. \tag{10.20}$$

The coordinates of the centre of mass in the zero momentum frame are therefore

$$X_k = J_{k4}/icM. \tag{10.21}$$

We now calculate $dJ_{ij}/dt = icdJ_{ij}/dx_4 = d/dx_4\iiint (x_i Z_{j4} - x_j Z_{i4}) dV$ and note that, since x_i is now merely one of the independent variables and not a particle coordinate, $dx_i/dx_4 = \delta_{i4}$. We thus obtain

$$dJ_{ij}/dt = \int (x_i \partial Z_{j4}/\partial x_4 + \delta_{i4} Z_{j4} - x_j \partial Z_{i4}/\partial x_4 - \delta_{j4} Z_{i4}) dx_1 dx_2 dx_3.$$

In an isolated system $\partial Z_{jk}/\partial x_k = 0$ and so $\partial Z_{j4}/\partial x_4 = -\partial Z_{j\lambda}/\partial x_\lambda$ where λ runs from 1 to 3. This leads to

$$x_i \partial Z_{j4}/\partial x_4 = -x_i \partial Z_{j\lambda}/\partial x_\lambda = -\partial/\partial x_\lambda (Z_{j\lambda} x_i) + \delta_{i\lambda} Z_{j\lambda}.$$

With a similar expression for $\partial Z_{i4}/\partial x_4$ the integrand contains $\delta_{i4} Z_{j4} + \delta_{i\lambda} Z_{j\lambda} = Z_{ji}$ and $-(\delta_{j4} Z_{i4} + \delta_{j\lambda} Z_{i\lambda}) = -Z_{ij}$ and, because the energy-momentum tensor is symmetric these two terms cancel.

This leaves two complete space derivatives and if the volume of integration contains the whole system these will give vanishing surface integrals. Thus we obtain the equation

$$dJ_{ij}/dt = 0 \tag{10.22}$$

whose spatial components express conservation of angular momentum and whose mixed space-time components give the Centre of Mass theorem

$$M\mathbf{X} = t\,\mathbf{P} + \text{constant.} \tag{10.23}$$

This equation is less trivial than it looks. It tells us that any isolated system, even if it has internal structure, behaves externally like a simple particle of inertial mass M at the position of the centre of mass. It also tells us how to specify the location of the system when only the external features of its motion are significant. It justifies basing the relativistic laws of dynamics on the behaviour of bodies in motion without regard to their internal structure, for it confirms that this procedure leads to a self-consistent theoretical scheme.

10.6 Orbital and Spin Angular Momentum

Classically the total angular momentum **J** of a system can be separated into its orbital angular momentum **L** associated with the motion of its centre of mass and its internal spin angular momentum **S** associated with rotation about the centre of mass. A Galilean transformation to moving coordinates alters **L** but not **S** and so the separation is invariant. **S** is also the part of **J** which is independent of the choice of origin (which again only affects **L**). We could define **S** as either the angular momentum about the centre of mass or as the origin independent part of **J**. Classically the two definitions are equivalent but, in relativity, although we can indeed give a covariant definition of spin as the origin independent part of the total angular momentum, this is not the same as the internal angular momentum about the centre of mass, except in the centre of mass frame, where in any case the spin and the total angular momentum are the same.

Whether J_{ij} is expressed in terms of the energy-momentum tensor as $J_{ij} = (i/ic)\iiint(x_iK_{j4}-x_jK_{i4})\,dV$ or as the sum over the particles $\sum(x_ip_j-x_jp_i)$, it is a covariant tensor and, in a new inertial frame with $x_i'=a_{ij}x_j$, its components will be

$J_{ij}' = a_{ik}a_{jl}J_{kl}$. The spatial components give the angular momentum and the mixed space-time components describe the motion of the centre of mass X_k as $MX_k - tP_k = (1/ic)J_{k4}$.

If the frame F^* is obtained from F by a translation of the origin, so that $x^*_i = x_i + a_i$, then the effect on J_{ij} is $J_{ij} \rightarrow J^*_{ij} = J_{ij} + (1/ic)\int(a_i K_{j4} - a_j K_{i4})dV = J_{ij} + a_i P_j - a_j P_i$ where P_i is the total momentum 4-vector. In the centre of mass frame $P_1 = P_2 = P_3 = 0$ and $P_4 = iM_o c$ and so in this particular frame the angular momentum components are unchanged, but for the other components we have $J^*_{k4} = J_{k4} + a_k P_4 = J_{k4} + iM_o c a_k$. Thus J_{ij} as a whole depends on the choice of the origin.

We would like to separate J_{ij} into $L_{ij} + S_{ij}$ where, in any frame, all the components of S_{ij} would be origin independent, and (so that it would coincide with the internal spin angular momentum in the rest frame F^R) its spatial components $S_{\lambda\mu}$ in the rest frame would equal $J_{\lambda\mu}$.

Consider the covariant second rank tensor

$$L_{ij} = -(1/M_o^2 c^2)[J_{ik}P_k P_j + J_{kj}P_k P_i]. \qquad (10.24)$$

Moving the origin by a_i alters J_{ij} to $J^*_{ij} = J_{ij} + a_i P_j - a_j P_i$, and L to $L^*_{ij} = L_{ij} - [(a_i P_k - a_k P_i)P_k P_j + (a_k P_j - a_j P_k)P_i P_k]/(M_o c)^2$. The terms containing $P_i P_j P_k$ cancel, thus leaving only $-P_k P_k(a_i P_j - a_j P_i)/(M_o c)^2$ and, since $P_k P_k = -(M_o c)^2$, this leaves

$$L^*_{ij} - L_{ij} = a_i P_j - a_j P_i = J^*_{ij} - J_{ij}, \qquad (10.25)$$

Therefore the second rank covariant tensor

$$S_{ij} = J_{ij} - L_{ij} = J_{ij} + [J_{ik}P_k P_j + P_i P_k J_{kj}]/(M_o c)^2 \qquad (10.26)$$

is origin independent. In the rest frame F^R, where the only non-zero component of P_i is $P^R_4 = iM_o c$, we have

$$L^R_{ij} = J^R_{ij}\delta_{j4} + \delta_{i4}J^R_{4j}. \qquad (10.27)$$

This has no purely spatial components and since its mixed space-time components are the same as those of J^R we see that $S^R = J^R - L^R$ has no mixed components and its purely spatial components are the same as those of J and are therefore the components of the internal spin angular momentum in the rest frame.

The tensor S can be constructed from J and P in any frame and although it is not the true internal spin angular momentum about the centre of mass, except in the centre of mass rest frame, it gives us a method of calculating the true spin in the rest frame from J and P in an arbitrary frame.

Consider next the true spin σ in an arbitrary frame. In the rest frame F^R the origin is at the centre of mass and at

t=0 coincides with the origin of F which moves along the X_1 axis of F^R with a velocity u. In this case $J^R{}_{4k} = -J^R{}_{4k} = 0$ and $L^R{}_{ij} = 0$. Since L is covariant it is zero in any frame obtained from F^R by a Lorentz transformation, thus in any of these frames F, the total angular momentum J about the origin 0 of F will be equal to S, the part of J that is origin independent. In classical, but not relativistic, dynamics this would imply that 0 was the centre of mass. At t=0 the centre of mass is at $X_k = J_{k4}/(icM)$ and, with F in motion along the X_1 axis, we will have $J_{14} = 0$, $J_{24} = -iu\gamma(u)J^R{}_{21}/c$, $J_{34} = -iu\gamma(u)J^R{}_{31}/c$, so that at t=0, $X_1 = 0$, $X_2 = uJ^R{}_{12}/(M_o c^2)$, $X_3 = -uJ^R{}_{31}/(M_o c^2)$. In F the system has momentum $P_1 = -\gamma M_o u$ and so the orbital angular momentum of the centre of mass about the origin has components $\ell_{12}=-X_2P_1=(u^2/c^2)\gamma J^R{}_{12}$, $\ell_{23}=0$, $\ell_{31}=X_3P_1=(u^2/c^2)\gamma J^R{}_{31}$. The true spin angular momentum σ has therefore components $\sigma_{23}=J_{23}$, $\sigma_{31}=J_{31}-\ell_{31}$, $\sigma_{12}=J_{12}-\ell_{12}$. The Lorentz transformation from F^R to F gives $J_{23}=J^R{}_{23}$, $J_{31}=\gamma J^R{}_{31}$ and $J_{12}=\gamma J^R{}_{12}$, thus the components of the true internal angular momentum about the centre of mass are

$$\sigma_{23}=J_{23}, \quad \sigma_{31}= \gamma J^R{}_{31}(1-u^2/c^2)=J^R{}_{31}/\gamma, \quad \sigma_{12}=\gamma J^R{}_{12}(1-u^2/c^2)=J^R{}_{12}/\gamma, \tag{10.28}$$

and the behaviour of σ is different from that of S.

As $u \to c$ the two transverse spin components $\sigma_2 = \sigma_{31}$ and $\sigma_3 = \sigma_{12}$ tend to zero as $1/\gamma(u)$, whereas the longitudinal component, $\sigma_1 = \sigma_{23}$ parallel to the velocity u, is unchanged. This contrasts with the transverse components of both the orbital and the total angular momentum L and J, which increase due to the increasing inertial mass and increasing distance of the centre of mass from the $0X_1$ axis.

There is one further aspect of these results which is of some interest. If the system has angular momentum **J** in its rest frame, with components $J_1=J_{23}$, $J_2=J_{31}$, $J_3=J_{12}$, we can write the transverse displacement of the centre of mass in the frame F as $X_2 = u_1J_3/M_oc^2$, $X_3 = -u_1J_2/M_oc^2$, and these will also be the transverse displacements of the centre of mass relative to its original position in the rest frame. These results are obviously part of a more general relation

$$\mathbf{X}^R = \mathbf{J}^R \times \mathbf{u}/M_oc^2. \tag{10.29a}$$

Thus if a system has a spin J in its rest frame and it is then considered in another inertial frame with any relative velocity up to c, the centre of mass will lie within a circle of radius

$$r = J^R/M_o c \tag{10.29b}$$

about the direction of $\mathbf{J}^R$. If the system has linear dimensions R and the maximum velocity within the system is v, then $J < M_o Rv$ and $r/R < v/c$. Usually r/R will be very small for macroscopic bodies, but consider a particle of spin $\hbar/2$ and rest mass m so that $r = \hbar/(2mc)$. The Compton wavelength is $\lambda = h/mc = 4\pi r$. We might also relate r to the classical Thomson radius a obtained by equating mc^2 and $e^2/8\pi\varepsilon_O a$ and the result is

$$r/a = 4\pi\varepsilon_o \hbar c/e^2 \approx 137, \tag{10.29c}$$

a result which certainly emphasises the essentially quantum mechanical nature of electron spin.

10.7 *Helicity*

There will be no rest frame for systems with zero rest mass and velocity c, such as an electromagnetic wave or a neutrino, for the object's velocity is c in all inertial frames. Nevertheless the angular momentum is a well defined quantity, part of a covariant tensor J_{ij}. We can compare the angular momenta in two frames related by a Lorentz transformation, and also calculate, in any frame in terms of J_{ij} and the 4-momentum p_i, the components of the internal spin angular momentum $\sigma_{\lambda\mu}$, even if these are not part of a covariant tensor. Also, if in a particular frame F the inertial mass is M and the maximum linear dimension of the system is R, none of the components of $\sigma_{\lambda\mu}$ can exceed $McR = -iP_4R$. Let J_{ij} be the angular momentum tensor in F, and suppose that in F the system has a momentum P_1 along the X_1 axis, then (since in any frame $P_jP_j=0$) we have $iMc=P_4=iP_1$. Now consider a frame F' with a velocity $u = \beta c$ in the same direction as the system. In this frame

$$P_1' = \gamma(u)(P_1+i\beta P_4) = \gamma(1-\beta)P_1 = [(1-\beta)/(1+\beta)]^{1/2}P_1,$$

$$P_4' = \gamma(u)(P_4-i\beta P_1) = \gamma(1-\beta)P_4 = [(1-\beta)/(1+\beta)]^{1/2}P_4.$$

As $\beta \to 1$ both P_1' and P_4' become small, but their ratio remains constant (because the velocity is c in both F' and F). The transformation of J gives

$$J_{23}' = J_{23}, \quad J_{31}' = \gamma(J_{31}+i\beta J_{34}), \quad J_{12}' = \gamma(J_{12}-i\beta J_{24}),$$

$$J_{14}' = J_{14}, \quad J_{24}' = \gamma(J_{24}+i\beta J_{12}), \quad J_{34}' = \gamma(J_{34}-i\beta J_{31}).$$

In F' the system still has $P_2' = P_3' = 0$ and the components X_2' and X_3' of the centre of mass coordinates are

$$X_2' = J_{24}'/P_4' = \gamma(J_{24}+i\beta J_{12})/P_4', \quad X_3' = J_{34}'/P_4' = \gamma(J_{34}-i\beta J_{31})/P_4'$$

so that the transverse components of the orbital angular momentum are

$$\ell_{31}' = X_3 P_1' = P_1'\gamma(J_{34}-i\beta J_{31})/P_4' = -i\gamma(J_{34}-i\beta J_{31}), \qquad (10.30a)$$

$$\ell_{12}' = -X_2 P_1' = -P_1'\gamma(J_{24}+i\beta J_{12})/P_4' = i\gamma(J_{24}+i\beta J_{12}), \qquad (10.30b)$$

and therefore the corresponding components of the internal angular momentum are

$$\sigma_{31}' = \gamma(J_{31}+i\beta J_{34}+iJ_{34}+\beta J_{31}) = \gamma(1+\beta)(J_{31}+iJ_{34}), \qquad (10.31a)$$

$$\sigma_{12}' = \gamma(J_{12}-i\beta J_{24}-iJ_{24}+\beta J_{12}) = \gamma(1+\beta)(J_{12}-iJ_{24}). \qquad (10.31b)$$

Both these components must be numerically less than $M'cR'$ where $M'=P_4'/ic$ and $R'$is the largest linear dimension in F'. Now $M'=P_4'/ic = [(1-\beta)/(1+\beta)]^{1/2}P_4/ic = [(1-\beta)/(1+\beta)]^{1/2}M$, and R', though it may be contracted to R in F, cannot be bigger than $R\gamma(u)$. Consider then the component σ_{31}', in which we require $\gamma(u)(1+\beta)(J_{31}+iJ_{34})<[(1-\beta)/(1+\beta)]^{1/2}\gamma McR$. If this is to be satisfied for all values of $\beta = u/c$ up to unity, it can only be because $J_{31}+iJ_{34} = 0$. Thus, in the original frame F

$$J_{34} = iJ_{31} \qquad (10.32a)$$

and similarly

$$J_{24} = -iJ_{12}. \qquad (10.32b)$$

In this case σ_{12}' and σ_{31}' will be zero in F', though there will be no restriction on the component of the spin σ_{23}' about the direction of motion.

Now consider σ_{12} and σ_{31} in the original frame F. We have $L_{31} = X_3 P_1 = P_1 J_{34}/P_4 = -iJ_{34} = J_{31}$, thus $\sigma_{31} = 0$ and similarly $\sigma_{12} = 0$. It follows that a system of zero rest mass can have no components of spin about directions normal to its direction of motion, though of course it may have orbital angular momentum (associated with its linear momentum and a particular choice of an origin not on the line of motion). The only possible <u>spin</u> is about the line of motion.

This all looks very plausible but in fact these results have no classical interpretation that ignores quantum effects. An electromagnetic wave packet of finite lateral extent in three dimensions travelling along a straight line with a velocity c is <u>not</u> a solution of Maxwell's equations. If the packet has a lateral extent R it contains waves with transverse wave numbers of order at least $K = 1/R$. But the field satisfies the wave equation so that its frequency f, and its wave number k in the direction of propagation, satisfy $(f/c)^2 = K^2 + k^2$. The product $(f/k)\mathrm{d}f/\mathrm{d}k$, of the phase velocity and the group velocity is c^2, but the phase velocity is greater than c and the group velocity is less. The relation between its momentum P and its energy E, or its mass $M = E/c^2$, is not $P = Mc$ or $P = E/c$ but $P = M\,\mathrm{d}f/\mathrm{d}k$ and we no longer have $P_1^2 + P_4^2 = 0$. There is a rest frame for the system and if, rather than a wave packet, we considered the complete radiation field of a source we would find the same result. Indeed, (see Eq.9.28) the rest frame of the radiation created by a single charged particle making a small change in its velocity is the rest frame of the particle. Thus, as far as electromagnetic waves go, these classical, but relativistic, results are somewhat unreal. We note that we could also have obtained similar results [see Eq.(10.28)] by considering the behaviour of a massive particle, as its velocity tends to c.

The results have, however, a direct and important application in quantum mechanics where the notion of a photon arises from the quantum description of measurement and of the state of a system. The photon corresponds to our hypothetical wave packet. When we detect energetic X-rays with a counter it is the photon, or particle, aspect of the rays, not their classical description as a wave, that is physically relevant. The classical relativistic theory tells us that in those circumstances the intrinsic angular momentum, if not zero, can only be about the photon's line of motion. There are of course other quantum constraints on the angular momentum of photons but the classical relativistic argument explains why, though quantum mechanics requires photons to have an intrinsic angular momentum $L\hbar$ with L=1, only two of the expected $2L$+1=3 states actually occur. The cosine of the angle between the axis of angular momentum and the photon linear momentum can only take the values ±1. These are referred to as the helicity of the photon. Other zero rest mass particles are the neutrinos with spin $\hbar/2$. These have a unique helicity +1, which associates angular momentum, an axial vector, with linear momentum, a polar vector, and is closely connected with parity non-conservation in weak interactions, and also perhaps with the solar neutrino problem.

These consequences are, however, quantum mechanical. The classical result is simply that relativity prevents us from describing a system as having zero rest mass and at the same time ascribing to it a spin angular momentum with components transverse to its momentum.

10.9 Discussion

Because force density **f**, unlike force **F** itself, forms the spatial part of a covariant 4-vector, the simplest generalisation of Newton's law, $\mathbf{F} = M\mathrm{d}\mathbf{v}/\mathrm{d}t = \mathrm{d}\mathbf{p}/\mathrm{d}t$, a generalisation which keeps the left hand side of the equation separate from the particular particle or particles on which the force acts, is based on considering $\mathbf{f}=\mathrm{d}\mathbf{q}/\mathrm{d}t$ where **q** is the momentum density. In the covariant expression of this relation we cannot replace t by the proper time, for this can only refer to a single body. We must therefore replace t by x_j and as a result **q** must be replaced by a covariant 2nd rank tensor. This leads us to the idea of an energy-momentum tensor as the natural tool to use in treating continuous media, fields or systems of particles interacting through fields. It also turns out to be well adapted to dealing with the motion of a single particle and its self-field. The energy-momentum tensor has a further very fundamental significance for it assigns covariant quantities to a composite system to describe its total mass, momentum, energy and angular momentum and thus justifies treating composite systems on the same footing as idealised particles with no internal structure.

It is sometimes straightforward to construct a suitable, symmetric energy momentum tensor to describe a system, and even to partition it in a natural way into two or more parts which are individually symmetric. This is, for example, the case for a system of charged particles in vacuum interacting through the electromagnetic field. In this case the total tensor is the sum of a particle tensor and a field tensor, but in other cases it may be much more difficult and then it is important not to apply to the separate parts arguments which strictly only apply to the total tensor. A typical example occurs for fields in a dielectric medium, not only is it impossible to decide from first principles whether terms involving the polarisation vector **P** belong in the material part or the field part, but one is also uneasily aware that even the elastic properties of the medium have an electromagnetic origin.

Further problems will arise in dissipative processes that involve heat as well as work and which will obviously be important in relativistic thermodynamics. Consider a body in internal equilibrium which, in a frame F, has a macroscopic momentum p_i and 4-velocity u_i. If heat Q' (measured in the rest frame) is added to the body its mass increases by Q'/c^2. The entropy S of a system in internal equilibrium is essentially the logarithm of the number of microscopic complexions compatible with the body's macroscopic specification (or the number of bits of information needed to specify one of these complexions uniquely). In any rational scheme it must be an invariant scalar and then, if heat is to be treated as part of a 4-vector, or perhaps a tensor of higher rank, the relation $Q = TdS$, which defines absolute temperature, suggests that temperature may not be a scalar. A body which in its rest frame has a temperature T' may have some other temperature T in a different frame. Formulating a consistent covariant thermodynamics is obviously not a trivial task.

For reasons which have their roots in the fundamental structure of both classical and quantum mechanics, spin and orbital angular momentum play a central role in classifying elementary particles and the states of nuclei and atoms. Whatever the principles of quantum mechanics may indicate, the theoretical structure must also be compatible with relativity and we have seen some of the problems which arise, mainly because the tensors, which most conveniently correspond to spin and orbital angular momentum, need not correspond to a separation of the total angular momentum into an internal part and an orbital part associated with the motion of the centre of mass.

Problems

10.1 In a fluid, with a classical velocity **v**, momentum density **q** and a classical energy density e, the basic equations obeyed by the fluid in the presence of a classical force density **f** are the momentum and energy balance equations $\partial q_\lambda v_\mu)/\partial r_\mu + \partial q_\lambda/\partial t = f_\lambda$ and $\partial(ev_\mu)/\partial r_\mu + \partial e/\partial t = \mathbf{f}.\mathbf{v} = f_\mu v_\mu$. If the fluid is compressible with density ρ and pressure π show that $\rho dv_\mu/dt = -\partial\pi/\partial r_\mu$

10.2 A perfect gas is a system of non-interacting particles in random motion. For simplicity we assume that they all have the same rest mass m. The average value of a component of the energy-momentum tensor K is $K_{ij} = n\langle p_i v_j\rangle$ where n is the total density of particles, whatever their momenta, and the average <> is over the statistical distribution, which we assume is isotropic so that off-diagonal terms vanish. Show that the trace of K, $(\Sigma_i K_{ii})$ is

$$n\langle \mathbf{p}.\mathbf{v} - E\rangle = n\langle \gamma m v^2 - \gamma mc^2\rangle = -n\,mc^2/\langle\gamma(v)\rangle,$$

and use this to find an expression for the pressure π. Give both the classical and extreme relativistic values of π.

CHAPTER 11

THE PRINCIPLE OF LEAST ACTION

Concepts associated with the principle of least action such as conjugate variables, the Hamiltonian and path integrals play an important role in classical and quantum mechanics; it therefore seemed sensible to end with a short treatment of the principle and how it is modified by relativity.

The instantaneous configuration of a dynamical system is specified by the values of a number of dynamical variables which we will denote by q_k with k = 1,2,3,4,5 etc. or q for short. They may be the 3N coordinates of N particles or the amplitudes of a rather small number of field components specified at each point in space, for example the configuration of the electromagnetic field can be specified by the values of the four components of the 4-potential A_j throughout space.

Successive configurations of a system of particles are specified by a succession of values for the particle coordinates, and we may regard these dynamical variables as functions of the single independent variable t, the time. Successive configurations of the electromagnetic field, on the other hand, can be described in terms of the four dynamical variables A_j which are functions of the four independent variables x,y,z and t, or x_i.

As time progresses configurations evolve according to some definite law, e.g. Maxwell's equations or Newton's laws. In all cases that we understand reasonably well, the equations of motion are second-order differential equations and therefore the succession of configurations, constituting the path or motion of the system, will depend on the initial velocities, or derivatives with respect to the independent variables, as well as the initial configuration. If we propose to regard the state of a system as implying not only its present configuration but all its future configurations as well, the state will depend on both the dynamic variables and their first derivatives with respect to the independent variables. This all suggests that we should be able to describe the behaviour of a system in terms of a function of the dynamical variables and their first derivatives.

Eighteenth and nineteenth century divines, philosophers, physicists and mathematicians, proceeding either by invoking supposed attributes of the Deity, by analogy or by investigation of known equations of motion, constructed several such theories. Their history (see e.g. Lindsay and Margenau 1936 or Yourgrau and Mandelstam 1968) is instructive but we will only be considering one particular theory, Hamilton's principle of least action, which has profoundly influenced modern physics.

If a single function is to determine the complete trajectory of a system it must make reference to the state of the system all along the trajectory. Let θ be a parameter, possibly but not necessarily the time t, such that each value of θ corresponds uniquely to each successive configuration. Next let $L(q,\dot{q})$ be a function of the dynamical variables q and their derivatives $\dot{q}=dq/d\theta$, and then finally form the integral $W = \int_a^b L d\theta$ from the initial state at $\theta = a$ to the final state at $\theta = b$. A given physical path will determine some particular value of W and we can regard W as a function of one initial state on the path and the path from that state to another final state.

There will be only one <u>physical</u> path starting with a given initial <u>state,</u> though there will be many <u>physical</u> paths starting from a given initial <u>configuration</u> <u>a</u>, but only one of these will lead from <u>a</u> to a given final <u>configuration</u> <u>b</u> and so the value of W will be determined when these two configurations are given. Thus instead of giving the <u>state</u> at $\theta = a$ (coordinates and velocities) to determine the path, we could give the <u>configurations</u> <u>a</u> and <u>b</u>(coordinates only) at a and b. Both approaches are fruitful but we will only consider the <u>action</u> $W = \int_a^b L d\theta$ with <u>a</u> and <u>b</u>, the initial and final <u>configurations</u> fixed.

The action has a definite value for a physical path, i.e. a path allowed by the laws of motion, from <u>a</u> to <u>b</u>. We can however, as a purely formal idea, consider paths from <u>a</u> to <u>b</u> which are not physical and do not satisfy the equations of motion. In particular we can consider paths which differ only very slightly from the physical path and consider the small difference ΔW that they make to the action integral.

If, for some reason, the integral along the physical path is an extremum, i.e. either a maximum or a minimum, this can be expressed as $\Delta W = 0$.

Hamilton's principle states that it is possible to find a function L (known as the Lagrangian) such that the conditions for ΔW to be zero are the equations of motion.

If we are only considering a single system it does not matter whether we take W to be a maximum on the physical path or a minimum. But if we are going to consider two or more systems, first separately and then interacting, we must adopt a definite convention, and we follow the historic precedent implicit in the name "Least Action" and take W to be a minimum on the physical path.

The generality of the brief statement $\Delta W = 0$ has a number of consequences. With different forms for L, the action principle can describe, amongst many other things, planetary motion, molecular vibrations, electron optics, electromagnetism, electrical networks and geometric optics. It also provides a basic component of the conceptual foundations of quantum mechanics and a fundamental tool in statistical mechanics.

Another aspect of the principle of least action is that if W is invariant under a transformation of the coordinates, then the complete system that it describes is also invariant under the same transformation. Conversely only a Lagrangian L that leads to an action W with the same invariance can describe the system. Since finding the appropriate form of Lagrangian for an unfamiliar type of system is largely a matter of guesswork and trial and error, this is a very useful feature that often eliminates all but a few possibilities. Thus in any dynamical system the action must be a scalar invariant under all the transformations of the Lorentz group, and similarly, if it is to describe the dynamics of a crystal, it must be invariant under all the operations of the crystal symmetry group. It is a mistake to think, however, that these constraints are enough to determine a unique Lagrangian. The literature of physics is littered with examples where Lagrangians with the right symmetry, but otherwise apparently drawn from a hat, have turned out to omit important terms or include incorrect terms leading to results conflicting with the experimental facts. Invariance restricts one's freedom of choice, but it does not lead infallibly to the correct Lagrangian. It does however have another extremely useful property: to each invariance of the action there corresponds a conservation law. In the Lorentz group, invariance under rotations of the axes corresponds to conservation of angular momentum, under translations of the origin to conservation of the three components of linear momentum, under translation of the origin of time to conservation of energy; and invariance under a change to a new moving inertial coordinate system corresponds to a constant velocity for the centre of mass.

The reader will recollect that it is just this type of invariance that forms the corner stone of the principle of special relativity.

The action for two separate non-interacting systems must obviously be the sum $W_1 + W_2$ of the actions for the two systems separately, and the equation $\Delta W = 0$ splits into two disconnected parts, $\Delta W_1 = 0$ involving only dynamical variables from system 1 and $\Delta W_2 = 0$ involving only those from system 2. If, however, the systems interact there must be terms in W that contain variables from both systems. The equations of motion for system 1 will result from considering variations ΔW due only to variations in the trajectory of system 1 but they will contain, as parameters, variables from system 2, and vice versa. These cross terms describe the influence of one system on the other and can be segregated in a single term W_{12} so that $W = W_1' + W_{12} + W_2'$, where W_1' contains only variables from system 1 and W_2' from system 2. If the systems are separated we require $W_1' \rightarrow W_1$, $W_2' \rightarrow W_2$ and $W_{12} \rightarrow 0$, but the only term containing variables from both systems, and so dependent on the separation of the two systems, is W_{12}. Therefore $W_1' = W_1$ and $W_2' = W_2$ and we can always write the action for two interacting systems as the sum of their separate actions and a single term W_{12} which describes both the effect of 1 on 2 and of 2 on 1. Furthermore, if system 2 is a prescribed system exerting impressed forces on system 1, the term W_{12} must describe the effect of these impressed forces on 1. Equally it must also describe the effect on 2 of impressed forces due to 1. In electrodynamics this more or less boils down to saying that the charge that determines the force exerted by an electric field on a charged particle is the same as the charge that determines the field produced by the body.

11.1 Lagrange's Equations

We now show how the action principle leads to equations of motion and we begin with a system of classical particles. We let the Lagrangian be $L(q,\dot{q},y)$, where q stands for all the dynamical variables, $\dot{q}$ for all their derivatives with respect to the independent variable t or θ , and y stands for any parameters in the system, such as the masses of particles or the components of impressed fields (possibly varying with time). We next let the independent variable be the time t, so that the action integral is $W = \int L dt$.

The Lagrangian is an explicit function of the variables q and $\dot{q}$, and these themselves are functions of the single independent variable t, though the parameters y may also depend on t. The total time derivative is therefore

$$dL/dt = \dot{q}\partial L/\partial q + \ddot{q}\partial L/\partial \dot{q} + \dot{y}\partial L/\partial y, \qquad (11.1)$$

where a sum is implied over all the variables q, the parameters y, and the time derivatives which are, since t is the only independent variable,

$$\dot{q}=\partial q/\partial t \equiv dq/dt, \quad \ddot{q}=\partial \dot{q}/\partial t \equiv d\dot{q}/dt \text{ and } \dot{y}=\partial y/\partial t \equiv dy/dt.$$

It will be convenient to regard the variation of L with t, that arises from q and $\dot{q}$, as an implicit variation but to treat the variation arising from y as explicit. We thus adopt the slightly artificial, though conventional, notation $\partial L/\partial t = \dot{y}\partial L/\partial y$ and write

$$dL/dt = \dot{q}\partial L/\partial q + \ddot{q}\partial L/\partial \dot{q} + \partial L/\partial t. \qquad (11.2)$$

The action integral W depends on the limits of integration, the form of L and the path, i.e. the behaviour of the functions q, $\dot{q}$ and y as functions of t. We now keep the limits fixed at t_a and t_b and, whatever the path from a to b, these initial and final configurations are kept unchanged.

The variation in W occurs as a result of small changes in the functional form of the variables $q(t)$ and $\dot{q}(t)$. We need not consider variations in the parameters $y(t)$ (which are to be regarded as impressed parameters). The key to our final result is that a small variation $\delta\dot{q}$ in a velocity is completely determined as

$$\delta\dot{q} = \partial(\delta q)/\partial t \equiv d(\delta q)/dt, \qquad (11.3)$$

by the variation δq in q. If, for example, $q=vt$ and $\delta q=t^2\delta a$ then $\delta\dot{q} = 2t\delta a$. For this type of variation

$$\Delta W = \int_a^b (\partial L/\partial q\delta q+\partial L/\partial\dot{q}\delta\dot{q})dt = \int_a^b \{\partial L/\partial q\delta q+\partial L/\partial\dot{q}d(\delta q)/dt\}dt$$

and the last term can be integrated by parts so that

$$\Delta W = \Big[\ \partial L/\partial\dot{q}\ \delta q\ \Big]_a^b + \int_a^b [\partial L/\partial q - d/dt(\partial L/\partial\dot{q})]\delta q\ dt.$$

Since the initial and final configurations a and b are not varied, δq in the integrated part is zero. Thus

$$\Delta W = \int_a^b [\partial L/\partial q - d/dt(\partial L/\partial\dot{q})]\delta q\ dt \qquad (11.4)$$

and now, since δq is an arbitrary function of t, $\Delta W = 0$ requires the coefficient of δq in the integrand to be everywhere zero. The Lagrangian L must therefore satisfy

$$\partial L/\partial q_k - d/dt(\partial L/\partial\dot{q}_k) = 0. \qquad (11.5)$$

These are *Lagrange's Equations* and there will be one such equation for each independent dynamical variable in the system. For example if there are N particles there are 3N equations.

With the total time derivative written out in full, the typical equation is

$$\partial L/\partial q_k - [\partial/\partial t+\dot{q}_j\partial/\partial q_j+\ddot{q}_j\partial/\partial\dot{q}_j]\partial L/\partial\dot{q}_k = 0 \qquad (11.6)$$

where a sum is implied over the other particle labels j, but not over k. We see that equation (11.6) is of second order in (d/dt) and therefore acceptable as the equation of motion for the dependent variables (particle coordinates) q_j.

As an illustration let $L = \frac{1}{2}m\dot{q}^2-V(q)$, then (11.6) becomes $\partial V/\partial q + m\ddot{q} = 0$ and, if V is a potential and q a Cartesian coordinate, this is obviously the equation of motion of a particle of mass m.

Provided that we have found the appropriate form of Lagrangian L, the action principle will lead to the correct equations of motion. If we are dealing with the motion of particles and if all the forces are conservative and can be expressed as the gradient of a potential V (which is independent of velocity), then the Lagrangian T-V, where T is the kinetic energy, will always lead to the appropriate equations of motion.

Unfortunately this excludes magnetic forces and so in electrodynamics, as in some other branches of physics, the discovery of a suitable Lagrangian is largely a matter of trial and error. Fortunately most of the trials and errors were made many years ago and, also fortunately, the correct Lagrangian is not unique, for if L leads to the correct equations of motion so will $L + dF/dt$ where F is an arbitrary function; and this gives one a better chance of stumbling across a Lagrangian that will do. For a proof of this result and an illuminating discussion of the basic structure of the action principle the reader could consult Rosen (1969). It is a mistake to believe that there is any formal way of constructing a Lagrangian.

The mathematical formulation of the action principle for a continuous system such as a fluid or a field will take a different form, because the dynamical variables are then functions of four independent variables x,y,z,t or x_i rather than just t alone. We must construct an action that refers to all points in space as well as all instants in time, and this leads us to consider a Lagrangian density L such that the action is

$$W = \int L\,dx\,dy\,dz\,dt = \int L\,dx^4/ic. \tag{11.7}$$

The Lagrangian density L will be an explicit function of the field variables q and their first derivatives. Since the context will usually tell us whether L is a Lagrangian or a Lagrangian density we will use the same letter and we will often ignore the factor $1/ic$ in relativistic calculations. When we bother to write out the label of a variable it will be as a *superscript*, thus q^k, and we will reserve *subscripts* to denote derivatives, thus $q_j \equiv \partial q/\partial x_j$ and $q_{ij} \equiv \partial^2 q/\partial x_i \partial x_j$. We note that these *partial* derivatives mean no more than that in evaluating, for example, $\partial q/\partial x_2$ we keep x_1, x_3 and x_4 constant. As far as the implicit and explicit appearance of the independent variables in L is concerned they are *total* derivatives. This ambivalence is neatly concealed by the notation $q^k{}_j$. We then have

$$L = L(q^k .., q^k{}_j .., Y), \tag{11.8}$$

where Y represents the various parameters of the system. The total derivative of L with respect to x_i at constant values of the other independent variables (which we will denote by L_i or dL/dx_i), is

$$L_i \equiv dL/dx_i = q^k{}_i \partial L/\partial q^k + q^k{}_{ij} \partial L/\partial q^k{}_j + (\partial L/\partial Y)(dY/dx_i) \tag{11.9}$$

with sums over both k and j.

The action integral is now to be taken over a path from one configuration a to another b, and the limits of 4-fold integration encompass the whole of 3-space as well as the time from a to b, i.e. a region of space-time.

For a variation of the trajectories

$\Delta W=\int[\delta q^k\partial L/\partial q^k+\delta q^k{}_j\partial L/\partial q^k{}_j]\,dx^4 =\int[\delta q^k\partial L/\partial q^k+\partial L/\partial q^k{}_j\ d\delta q^k/dx_j]\,dx^4.$

A partial integration yields a term $\left[\delta q^k\ \partial L/\partial q^k{}_j\ \right]_a^b$ which is zero because the variation vanishes at the end points and this leaves

$$\Delta W = \int[\partial L/\partial q^k - d/dx_j(\partial L/\partial q^k{}_j]\delta q^k dx^4.$$

For this to be zero for arbitrary variations δq^k requires

$$\partial L/\partial q^k - d/dx_j(\partial L/\partial q^k{}_j) = 0. \qquad (11.10)$$

These equations, one for each dynamical variable labelled by k, are the dynamical, Lagrange, field equations. In the form

$$\partial L/\partial q^k -[q^k{}_i\partial/\partial q^k+q^k{}_{ij}\partial/\partial q^k{}_j+(dY/dx_j)(\partial/\partial Y)]L = 0, \qquad (11.11)$$

it can be seen that they are second order differential equations to be satisfied by the dynamical field variables q^k. Later on we will show how the electrodynamic equations can be expressed by an action principle, but first we look at the relativistic invariance of the principle.

11.2 The Invariant Action Principle

For a system of particles the action integral extends from one time and configuration to another time and another configuration. Its limits are therefore defined by two sets of events in which the particles are at prescribed positions at prescribed times and this is a relativistically invariant definition. Similarly, for a continuum, the region of integration in $W=\int L dx^4$ is a region of space-time, which is also a relativistically invariant concept and so, if W is a relativistically invariant quantity, the action principle is itself invariant, but it only leads to the right number of equations if W is a scalar. Thus for the action principle to be useful the action must be an *invariant scalar*. When $W=\int L dt$, $L dt$ must be a scalar but when $W=\int L dx^4$ the 4-volume of integration is a scalar and so the Lagrangian density itself must be a scalar. These considerations provide useful hints about constructing Lagrangians. For example, they suggest that in relating electromagnetism to the action principle we might first consider the suitability of the two simple scalar invariants $\frac{1}{2}(\varepsilon_0 E^2 - B^2/\mu_0)$ and **E.B**, before considering more complicated possibilities.

11.3 The Relativistic Free Particle

We seek a Lagrangian such that Ldt is a scalar and we also want L to approach the classical form $\frac{1}{2}m_0 v^2$ when $v<<c$. Because the particle is free $dW = \int Ldt$ cannot depend on the absolute values of the coordinates x_i but only on their differentials. The simplest possibility is $(dx_j dx_j)^{1/2}$ or some multiple of this quantity, and we try

$$Ldt = -m_0 c(-dx_j dx_j)^{1/2}. \qquad (11.12)$$

If $\mathbf{x}=\mathbf{v}t$ then $Ldt = -m_0 c(c^2-v^2)^{1/2}dt = -m_0 c^2(1-v^2/c^2)^{1/2}dt$, and if $v<<c$, this reduces to $L = -m_0 c^2 + \frac{1}{2}m_0 v^2$ which, apart from the irrelevant term $-m_0 c^2$, gives the classical form $\frac{1}{2}m_0 v^2$.

A possible trial form for the action is therefore

$$W = \int dW = \int -m_0 c(-dx_j dx_j)^{1/2} . \qquad (11.13)$$

Instead of identifing elements of the path using t we could use the proper time τ, and write (11.13) as

$$dW = -m_0 c(-dx_j/d\tau \, dx_j/d\tau \,)^{1/2} d\tau \qquad (11.14)$$

and then W can be expressed in terms of the 4-velocity

$$u_j = dx_j/d\tau = \gamma(v)(\mathbf{v}, ic) \qquad (11.15)$$

as

$$W = \int -m_0 c(-u_j u_j)^{1/2} d\tau. \qquad (11.16)$$

The principle $\Delta W = 0$ applied to (11.16) leads to the Lagrange equation (11.5) which is $d/d\tau(\partial L/\partial u_i)=0$, and this gives $d/d\tau[m_0 c u_i/(-u_j u_j)^{1/2}]=0$. Since $(-u_j u_j)^{1/2}=\gamma(v)(c^2-v^2)^{1/2} = c$ and m_0 is a constant, we finally obtain Newton's first law of motion in the form $du_i/d\tau = 0$. There are many other ways in which we can manipulate (11.13) and get rather more interesting results. The reader will find several good examples in Podolsky and Kunz (1969).

11.4 A Relativistic Particle in an Impressed Field

The action $dW = -m_0 c^2 d\tau = -m_0 c(-dx_i dx_i)^{1/2} = -m_0 c^2 dt/\gamma(v)$ leads to straight line trajectories but we could cause it to lead to trajectories of any form by replacing $(cd\tau)^2=-dx_j dx_j$

by $c^2(d\tau)^2 = -g_{ij}(x)dx_i dx_j$ and considering an action integral $W = \int(-g_{ij}(x)dx_i dx_j)^{1/2}$. This is the route usually taken in the treatment of gravitation in general relativity. It is generally described as incorporating the gravitational field into the coordinate system and it works because gravity acts equally on everything. An excellent treatment of the role of the principle of least action in general relativity is given by Dicke (1965).

Here, however, we shall only consider a particle of charge e in an electromagnetic field described by a 4-vector $A_i = (\mathbf{A}, i\phi/c)$, where $\mathbf{A}$ is the vector and ϕ the scalar potential. The quantity $eA_j dx_j$ is a scalar invariant and could possibly be part of the action; moreover when A_j is constant (so that the fields $\mathbf{E}$ and $\mathbf{B}$ are zero), $eA_j dx_j$ is a complete differential and adding it to the action has no effect on the Lagrange equations. We therefore try

$$dW = -m_o c^2 d\tau + eA_j dx_j \tag{11.17a}$$

and, since $dW = L dt$, the Lagrangian L will then be

$$L = -m_o c^2/\gamma(v) + eA_j dx_j/dt = -m_o c^2/\gamma(v) + e\mathbf{A}.\mathbf{v} - e\phi\,. \tag{11.17b}$$

The Lagrange equations for each component are

$$e\partial\phi/\partial x_\lambda - ev_\mu \partial A_\mu/\partial x_\lambda - d(mv_\lambda + eA_\lambda)/dt = 0$$

where $m \equiv m_o\gamma(v)$, λ and μ run from 1 to 3 and there is a sum over μ. Now $dA_\lambda/dt = \partial A_\lambda/\partial t + v_\mu \partial A_\lambda/\partial x_\mu$ and so we obtain

$$d(mv_\lambda)/dt = -e(\partial\phi/\partial x_\lambda + \partial A_\lambda/\partial t) + ev_\mu(\partial A_\mu/\partial x_\lambda - \partial A_\lambda/\partial x_\mu)\,.$$

The first term in brackets is $-E_\lambda$ and the second is the λth component of $\mathbf{v}\times(\nabla\times\mathbf{A}) = \mathbf{v}\times\mathbf{B}$ and so the Lagrange equation is the Lorentz equation $d(m\mathbf{v})/dt \equiv d(m_o\gamma(v)\mathbf{v})/dt = e\mathbf{E} + e\mathbf{v}\times\mathbf{B}$. The term $eA_j dx_j$ in the action (11.17a) is identical with the corresponding term in the classical non-relativistic action, and since in both cases it leads to the correct equation of motion, it is a proper choice.

11.5 The Electromagnetic Field

We now try to find an action principle that will lead to Maxwell's equations, $\nabla\times\mathbf{E} + \dot{\mathbf{B}} = 0$, $\nabla\times\mathbf{B} - \mu_o\varepsilon_o\dot{\mathbf{E}} = \mu_o\mathbf{j}$, $\nabla.\mathbf{B} = 0$, $\nabla.\mathbf{E} = \rho/\varepsilon_o$. At the very outset we are faced with a problem, since the action principle gives only as many equations as there are field variables, and there are only six components of $\mathbf{E}$ and $\mathbf{B}$ but altogether eight component Maxwell equations. Moreover the Lagrange equations are second order whereas the

Maxwell equations are first order. Initially we avoid this difficulty by using the components of the 4-potential A_i as the set of dynamical variables. With $\mathbf{B} = \nabla\times\mathbf{A}$ and $\mathbf{E} = -\dot{\mathbf{A}}-\nabla\phi$ this reduces the two homogeneous Maxwell equations $\nabla.\mathbf{B} = 0$ and $\nabla\times\mathbf{E}+\dot{\mathbf{B}} = 0$ to the status of identities, and leaves the two inhomogeneous equations with four components equal in number to the 4 components of A_j. This however leaves us with the difficulty that the potentials are arbitrary to within a gauge transformation $A_j \rightarrow A_j+\partial\psi/\partial x_j$, since this leaves the fields unchanged. We could perhaps avoid this particular difficulty by specifying a gauge in a Lorentz invariant way, but it is easier to eliminate it entirely by first constructing the action in terms of the fields alone, so that it will automatically be gauge invariant, and only later convert it to a formula in terms of A. We still have to find a Lagrangian, or rather a Lagrangian density, L. Because $W = \int L dx^4$ we require a scalar invariant for L, and we know two simple invariants, $\frac{1}{2}(\varepsilon_o E^2-B^2/\mu_o)$ and $\mathbf{E}.\mathbf{B}$. The choice $\mathbf{E}.\mathbf{B}$ is useless as it vanishes for two important cases, the electrostatic field and the magnetostatic field.

Omitting, for convenience, ε_o and μ_o we consider, as a possible Lagrangian density,

$$L = \tfrac{1}{2}(E^2-B^2) = -\tfrac{1}{4}F_{ij}F_{ij} = \tfrac{1}{4}F_{ij}F_{ji} = -\tfrac{1}{4}(\partial A_i/\partial x_j-\partial A_j/\partial x_i)^2, \quad (11.18)$$

where F_{ij} is the field tensor. However, because this makes no reference to ρ or $\mathbf{j}$, it cannot lead to the inhomogeneous field equations and so we must also find a term to describe the interaction of the field with the charges. In (11.17b) we used $eA_j dx_j/dt$ for a Lagrangian, now we try J_iA_i, where $J_i=(\mathbf{j},ic\rho)$. The total trial Lagrangian density is then

$$L = \tfrac{1}{2}(E^2-B^2)+ J_iA_i. \quad (11.19a)$$

giving an action

$$W = \int [\tfrac{1}{2}(E^2-B^2)+J_iA_i]\,dx^4. \quad (11.19b)$$

The field term is gauge invariant and thus changing A_i to $A_i + \partial\psi/\partial x_i$ only gives an additional contribution from J_iA_i, $\Delta_g W = \int J_i\partial\psi/\partial x_i dx^4 = \int[\partial/\partial x_i(J_i\psi) -\psi\partial J_i/\partial x_i]dx^4$. Because charge is conserved $\partial J_i/\partial x_i = 0$ and, since the remaining term in the integrand is a complete derivative, we see that $\Delta_g W$ can have no effect on the Lagrange equations derived from (11.19b). Again we notice the connection between charge conservation and gauge invariance.

We now write W in terms of the 4-potential as

$$W = \int [J_i A_i - \tfrac{1}{4}(\partial A_i/\partial x_j - \partial A_j/\partial x_i)^2]\,dx^4 \qquad (11.19c)$$

and consider the Lagrange equations obtained by varying the components of the potential (but not those of J_i). We need

$$\partial L/\partial q^k{}_i = \partial L/\partial(\partial A_k/\partial x_i) = -(\partial A_k/\partial x_i - \partial A_i/\partial x_k)\,. \qquad (11.19d)$$

The factor ¼ disappears, for each partial derivative occurs once in $\partial A_k/\partial x_i - \partial A_i/\partial x_k$ as the first and once as the second term, and also because we are differentiating a square. Thus with $\partial L/\partial A_i = J_i$, the equations of motion are

$$J_i + d/dx_j(\partial A_i/\partial x_j - \partial A_j/\partial x_i) = 0. \qquad (11.19e)$$

The component with i=1 is

$$\mu_o J_1 - \partial B_3/\partial x_2 + \partial B_2/\partial x_3 + (1/c^2)\partial E_1/\partial t = 0,$$

where we have restored the factor μ_o, and so we recognise the spatial components of (11.19e) as $\nabla\times\mathbf{B} - \mu_o\varepsilon_o\partial\mathbf{E}/\partial t = \mu_o\mathbf{j}$. The fourth component can easily be checked to be $\nabla.\mathbf{E} = \rho/\varepsilon_o$. Therefore the action (11.19b) leads to the correct field equations.

We note that Eq.(11.19e) can also be expressed as

$$(\nabla^2 - \mu_o\varepsilon_o\partial^2/\partial t^2)A_i = -\mu_o J_i + \partial/\partial x_i(\partial A_j/\partial x_j) \qquad (11.19f)$$

and, in the Lorentz gauge with $\partial A_j/\partial x_j = 0$, this becomes the familiar inhomogeneous wave equation.

11.6 Particles and Electromagnetic Fields

We could, by introducing the Dirac delta function, treat particles in terms of a Lagrangian density and thus combine the action principle for particles with that for fields in one equation. Fortunately this is unnecessary and we can always deal with the effects of fields on the particles using the Lagrangian $eA_j dx_j/dt$, and the effects of the particles on the fields using the Lagrangian density $J_i A_i$.

11.7 The Hamiltonian

We now leave Lagrange's second order equations and look at Hamilton's first order equations of motion, which are important in both classical and quantum mechanics.

The classical Lagrangian of a free particle is $L=\tfrac{1}{2}m_o\dot{q}_k{}^2$ and so its momentum is $p_k = m_o\dot{q}_k = \partial L/\partial\dot{q}_k$. We now adopt

$$p_k = \partial L/\partial\dot{q}_k \qquad (11.20)$$

as the general definition of the momentum p_k canonically conjugate to the variable q_k, though, of course, this definition will only be valid (or useful) if we have already constructed a Lagrangian L that leads to the correct equations of motion. We emphasise that, if L describes a system with velocity dependent forces, p will not coincide with the usual momentum $m\dot{q}$.

Next consider the function

$$H = \Sigma p_k \dot{q}_k - L(q_k, \dot{q}_k, y) = \Sigma \dot{q}_k \partial L/\partial \dot{q}_k - L(q_k, \dot{q}_k, y) \qquad (11.21)$$

which, in this form, is an explicit function of the variables $q_k, \dot{q}_k$ and the parameters y, so that its total differential is

$$dH = \textstyle\sum_k [p_k d\dot{q}_k + \dot{q}_k dp_k - (\partial L/\partial q_k) dq_k - (\partial L/\partial \dot{q}_k) d\dot{q}_k] - (\partial L/\partial y) dy.$$

By virtue of the definition (11.20), the first and the last terms in the sum cancel, leaving

$$dH = \Sigma[\dot{q}_k dp_k - (\partial L/\partial q_k) dq_k] - (\partial L/\partial y) dy. \qquad (11.22)$$

Now (11.20) consists of as many equations as there are velocity components $\dot{q}_k$ and we can use these equations (at least in principle, though luckily it is often quite easy) to express the velocities $\dot{q}_k$ in terms of the variables q_k and their conjugate momenta p_k. These results can then be used to express H as a function of the momenta p, the variables q and the parameters y, with the velocities eliminated. Expressed in this way as $H(p,q,y)$ the function is called the Hamiltonian, and its total differential is

$$dH = \textstyle\sum_k [(\partial H/\partial p_k) dp_k + (\partial H/\partial q_k) dq_k] + (\partial H/\partial y) dy. \qquad (11.23)$$

We have now only to compare (11.22) and (11.23) and use the Lagrange equation (11.10), to find $\partial H/\partial p_k = dq_k/dt$ and also $\partial H/\partial q_k = -\partial L/\partial q_k = -d(\partial L/\partial \dot{q}_k)/dt = -dp_k/dt$. Thus we obtain Hamilton's set of first order equations of motion:

$$dq_k/dt = \partial H/\partial p_k, \qquad (11.24a)$$

$$dp_k/dt = -\partial H/\partial q_k, \qquad (11.24b)$$

$$\partial H/\partial y = \partial L/\partial y. \qquad (11.25)$$

For a system of N particles the 6N first order Hamilton equations replace the 3N second order Lagrange equations. These first order equations are sometimes called the canonical equations of motion.

The total time derivative of H is

$$dH/dt = [\dot{p}_k \partial H/\partial p_k + \dot{q}_k \partial H/\partial q_k] + (\partial H/\partial y) dy/dt = (\partial H/\partial y) dy/dt \qquad (11.26)$$

and so, if the parameters y are constant, H is a constant of the motion. The total time derivative of the total momentum $\mathbf{P}$ is $dP_\alpha/dt = \Sigma d\mathbf{p}_\alpha/dt = -\Sigma\partial H/\partial q_\alpha$ where $\alpha = 1,2$ or 3 is a vector subscript. If the system is isolated not only will dy/dt be zero, but H cannot depend on the absolute position of the origin, and if the origin is translated by $d\mathbf{r}$ we have $\Sigma(\partial H/\partial q_\beta)dr_\beta = dH = 0$ and since $d\mathbf{r}$ is arbitrary and the same for all the particles we get $\partial H/\partial q_\alpha = 0$ and thus $d\mathbf{P}/dt = 0$. For an isolated system both the canonical momentum $\mathbf{P}$ and the Hamiltonian H are constant. A similar calculation based on invariance under a rotation of the axes shows that the total angular momentum $\mathbf{L} = \Sigma\mathbf{r}\times\mathbf{p}$ is also constant (see e.g. Landau and Lifshitz 1960).

The Hamiltonian formulation of dynamics is a restatement of the principle of Least Action and so this new formulation must be compatible with relativity if we start with a suitably invariant action function. However, because it treats time on a different footing from space, it is not manifestly covariant. It is therefore difficult to transform the equations in Hamiltonian form from one Lorentz frame to another. But if, starting with the action principle, we can obtain a Hamiltonian appropriate to a particular frame we may, within this frame, use the Hamiltonian to treat relativistic problems.

For a relativistic particle in an electromagnetic field described by the 4-potential A_i, the Lagrangian is

$$L = -m_o c^2/\gamma(v) + e\mathbf{A}.\mathbf{v} - e\phi, \qquad (11.27)$$

and so the canonical momentum has components $p_i = \partial L/\partial v_i = mv_i\gamma(v) + eA_i$, or

$$\mathbf{p} = m_o\mathbf{v}\gamma(v) + e\mathbf{A}, \qquad (11.28)$$

and this leads to

$$H = \mathbf{p}.\mathbf{v} - L = mc^2\gamma(v) + e\phi. \qquad (11.29)$$

When v has been eliminated this will be the Hamiltonian. Since $(\mathbf{p}-e\mathbf{A}).(\mathbf{p}-e\mathbf{A}) = [m_o\gamma(v)]^2 = (m_o c)^2(\gamma^2-1)$ we obtain

$$m_o c^2\gamma(v) = m_o c^2[1+(\mathbf{p}-e\mathbf{A}).(\mathbf{p}-e\mathbf{A})/(m_o c)^2]^{\frac{1}{2}}, \qquad (11.30a)$$

or

$$H = m_o c^2[1+|\mathbf{p}-e\mathbf{A}|^2/(m_o c)^2]^{\frac{1}{2}} + e\phi. \qquad (11.30b)$$

We notice that H in the form (11.29) is obviously the energy and also that it does not contain the vector potential, but when H has been expressed in canonical form in terms of the canonical momentum $\mathbf{p}$, the vector potential $\mathbf{A}$ reappears, as it must, to give magnetic terms in the canonical equations of motion. We leave it to the reader to verify that these equations, which are $\dot{q}_i = \partial H/\partial p_i$ and $\dot{p}_i = -\partial H/\partial q_i$, yield

$$d\mathbf{q}/dt = (\mathbf{p}-e\mathbf{A})/m_o\gamma, \quad dp_i/dt = ev_\mu\partial A_\mu/\partial q_i - e\partial\phi/\partial q_i. \tag{11.31}$$

The spatial part of the second equation is equivalent to $d[m_o\mathbf{v}\gamma(v)]/dt = e\mathbf{E} + e\mathbf{v}\times\mathbf{B}$.

When the velocities are small $\mathbf{p}-e\mathbf{A} = m\mathbf{v}\gamma(v) << mc$ and

$$H \to m_oc^2 + |\mathbf{p}-e\mathbf{A}|^2/2m_o + e\phi. \tag{11.32}$$

Apart from the term m_oc^2 this is the usual classical Hamiltonian.

11.8 *Conservation Laws*

The systematic investigation of the conservation laws associated with invariant properties of the action is due to E. Noether (1918) but we will not present her results in all their generality. We saw above that invariance of the action under translations of the origin in time and space implies conservation of energy and linear momentum. Similarly, invariance under rotations implies conservation of angular momentum, and invariance under a change from one frame of reference to another moving with a constant relative velocity implies a constant velocity for the centre of mass. A rather different result is that gauge invariance implies charge conservation. It is tempting to suppose that these various conservation laws are the consequences of the various invariances of the action, and thus of the fundamental properties of space and time expressed by the principle of relativity. We should, however, remember that we could only define inertial mass and construct either classical or relativistic dynamics, by insisting that the law of conservation of momentum was fundamental. Equally the displacement current which links electromagnetism to the velocity of light and leads to gauge invariance is totally dependent on charge conservation. Thus it is not surprising if a theory based on conservation laws leads to a restatement of those laws.

Nevertheless, the relation between conservation laws and invariances of the action is often useful in constructing trial theories in branches of physics with a less well verified experimental foundation than electromagnetism.

11.9 Relativity, Quantum Mechanics and Electron Spin

The non-relativistic Schrödinger equation for a free electron is obtained from the relation $H=E=p^2/2m$ by replacing H by $-i\hbar\partial/\partial t$ and $\mathbf{p}$ by $i\hbar\nabla$. The same procedure applied to $E^2=p^2c^2+m^2c^4$ yields the Klein-Gordon equation

$$\hbar^2\{\partial^2\psi/\partial t^2 - c^2\nabla^2\psi\} + m^2c^4 = 0. \quad (11.33)$$

This violates a fundamental principle of quantum mechanics (see Dirac 1958) because it is of second order in $\partial/\partial t$. If, to overcome this, the same prescription is applied to $E=c\{p^2+m^2c^2\}^{½}$ this will violate relativity by not treating $\partial/\partial t$ and ∇ in the same way, and it also yields a useless irrational expression containing ∇.

The second order wave equations of physics are almost always related to two or more first order equations. These may involve wave functions with more components than the function in the second order equation. Moreover, the variables in the first order equations are often needed to construct expressions corresponding to the density or flux of a conserved quantity, such as charge or energy. For example the energy density $½(\varepsilon_o E^2+\mu_o H^2)$ and Poynting's vector $\mathbf{E}\times\mathbf{H}$ are assembled from the six variables in Maxwell's equations, <u>not</u> directly from the potential A_i that satisfies $\Box^2 A_i=0$. These ideas are the basis of Dirac's solution to the problem of how to formulate a relativistic wave equation.

Dirac starts with the linear and rational equation

$$i\hbar\{c^{-1}\partial/\partial t - \alpha_1\partial/\partial x_1 - \alpha_2\partial/\partial x_2 - \alpha_3\partial/\partial x_3 - \beta\}\psi = 0 \quad (11.34)$$

and then insists that operating on this from the left with $i\hbar\{c^{-1}\partial/\partial t + \alpha_1\partial/\partial x_1 + \alpha_2\partial/\partial x_2 + \alpha_3\partial/\partial x_3 + \beta\}$ shall yield the Klein-Gordon equation (11.33). All the cross products must anticommute and also $\alpha_1{}^2=\alpha_2{}^2=\alpha_3{}^2=1$ and $\beta^2=m^2c^2$. Thus there will be six relations $\alpha_1\alpha_2+\alpha_2\alpha_1=0$, $\alpha_2\alpha_3+\alpha_3\alpha_2=0$, $\alpha_3\alpha_1+\alpha_1\alpha_3=0$, $\alpha_1\beta+\beta\alpha_1=0$, $\alpha_2\beta+\beta\alpha_2=0$ and $\alpha_3\beta+\beta\alpha_3=0$. The α's and β can only satisfy these relations if they are at least 4×4 matrices and this then implies that the wave function ψ will have four components. These components must correspond to internal properties of the electron. Dirac then shows that the matrices α and β can be chosen so that Eq. (11.34) yields a Lorentz invariant theory and he identifies the internal properties as the electron's spin $\hbar/2$ with two values ↑ and ↓, together with the sign of its charge.

Dirac further shows that these results will apply to any entity whose *position* is an observable, i.e. something which in the classical limit is a particle. The results do not apply to photons, the quanta of the electromagnetic field, or to the various quanta of the weak and strong forces, or to gravity.

The relativistic requirement that the electron has two spin states determines the form of the periodic table of the elements. If it were not for spin, carbon would be a very different sort of element.

11.10 *Discussion*

The Principle of Least Action yields an elegant and compact formulation of a physical theory or problem as long as we can find a Lagrangian, or Lagrangian density that yields the correct dynamical equations as its Lagrange equations. In the last section we have seen how the search for a relativistic quantum Hamiltonian leads to an explanation of the existence of electron spin. With all its implications for chemistry and biology this must surely be amongst the most important consequences of special relativity.

The action integral constructed from a proper Lagrangian must be an invariant scalar under all those transformations that leave the physical situation invariant. If a suitable and invariant action integral can be found, this guarantees that the dynamical equations had the correct invariance. Alternatively, when constructing a new theory, where we do not already know the dynamical equations, we can search for a suitable Lagrangian by requiring specific invariances, and also try to construct a Lagrangian as similar as possible to that in an already existing and successful theory. Thus quantum chromodynamics, the theory of the strong nuclear force, is based on quantum electrodynamics, itself derived from classical electrodynamics by Dirac's formal quantising procedures applied to the classical action principle.

The action principle yields pairs of canonically conjugate variables, q and $p = \partial L/\partial \dot{q}$ which, in quantum mechanics, obey commutation rules and are associated with uncertainty relations such as $\langle\Delta p^2\rangle\langle\Delta q^2\rangle \geq \hbar^2/4$. This method of finding the proper dynamical variables is essential in constructing the quantum mechanical generalisation of a classical theory, especially where the choice of conjugate variables is not immediately obvious (see e.g. Robinson 1965). The canonically conjugate variables are also needed to construct the Hamiltonian which plays a central role in quantum and statistical mechanics, and is a useful tool in solving specific problems. Finally the action principle may lead to the discovery of constants of the motion and conservation laws related to the invariances of the actual system.

Despite its elegance, we should not assume that all physical phenomena can necessarily be described in terms of an action principle, or that this will always be useful. Nor should we believe that there is **any** procedure, except trial and error, for constructing a valid action integral. Further, we cannot assume that a theory is automatically valid because it has been derived from a plausible, and suitably invariant, action integral. Although this is an excellent way of constructing trial theories whose consequences can be compared with experiment, there is certainly no guarantee of success. Perhaps the best way to approach the action principle is to regard it as a versatile, refined and very effective form of book-keeping.

Problems

11.1 A tuned circuit consists of an inductance L in parallel with a capacitance C, and the charge on C at any instant is q. If the Lagrangian for this system is $\frac{1}{2}L\dot{q}^2-\frac{1}{2}q^2/C$, find the Lagrange equation of motion, the Hamiltonian and the Hamilton equations of motion.

11.2 A possible Lagrangian for a free relativistic particle of rest mass m_0 and velocity $\mathbf{v}=d\mathbf{q}/dt$ is $L = -m_0c^2(1-v^2/c^2)^{\frac{1}{2}}$. Find the momentum conjugate to $\mathbf{q}$ and form the Hamiltonian.

11.3 Eq.(11.31) has $dp_4/dt=ev_\mu\partial A_\mu/\partial q_4-e\partial\phi/\partial q_4$ as its fourth component. Express this equation in terms of the energy of the particle and the fields.

APPENDIX

A1 Volume

In three dimensions the volume element is $dr_1dr_2dr_3$ and it is obvious from geometrical considerations that this is invariant under a rotation of the axes. On the other hand a surface element dr_1dr_2 is associated with a direction, its outward normal, and we do not expect it to remain invariant. Thus in four dimensions we expect to find that $dx_1dx_2dx_3dx_4$ is an invariant, but not the usual volume element $dx_1dx_2dx_3$, which in four dimensions is a "3-surface". We cannot now appeal to geometry to show that $dx_1dx_2dx_3dx_4$ is an invariant but must derive this property from the matrix a_{ij}.

The proof in four dimensions is tedious and so we indicate the method by considering how it would proceed in three dimensions.

Let $f(r_\lambda) = f(r_\lambda')$ be a scalar function of the variables r_λ or r_λ' and consider the integral

$$I = \int\int\int f(r_\lambda')dr_1'dr_2'dr_3'$$

which we intend to express as an integral over the variables r_λ. During the first integration over r_1' we keep r_2' and r_3' constant and so, during this integration $a_{1\mu}dr_\mu = dr_1'$, $a_{2\mu}dr_\mu = 0$ and $a_{3\mu}dr_\mu = 0$. We can solve these equations, in terms of the determinant $|a|$ of a, and of the cofactor

$$A_{11} = \begin{vmatrix} a_{22} & a_{23} \\ a_{32} & a_{33} \end{vmatrix}$$

and we obtain $|a|dr_1 = A_{11}dr_1'$. Thus we now have $I=\iiint(f\,|a|/A_{11})\,dr_1dr_2'dr_3'$. Next we integrate over dr_2' keeping r_1 and r_3' constant which requires $a_{22}dr_2+a_{23}dr_3 = dr_2'$ and $a_{32}dr_2+a_{33}dr_3=0$, so that $A_{11}dr_2 = a_{33}dr_2'$ and the integral is now $I=\int\int\int(|a|/a_{33})f\,dr_1dr_2dr_3'$. The final step is the integration over r_3' with r_1 and r_2 constant. This requires $a_{33}dr_3=dr_3'$, and so $I = |a|\int\int\int f(r_\lambda)dr_1dr_2dr_3$ and therefore, since $|a|=1$, $dr_1'dr_2'dr_3' = |a|dr_1dr_2dr_3 = dr_1dr_2dr_3$.

In general, if there are two sets of independent variables $z_1..z_n$ and $z_1'..z_n'$ related by $z_j'=a_{jk}z_k$ then $dz_1'..dz_n' = |a|dz_1..dz_n$. Since we are primarily interested in Lorentz transformations with n=4 and $|a|=1$, we will have

$$dx_1'dx_2'dx_3'dx_4' = dx_1dx_2dx_3dx_4. \qquad (A1.1)$$

Furthermore, since $dx_4 = ict$, we see that dxdydzdt must be an invariant scalar.

A2 *Force and Force Density*

Classically force $\mathbf{F}=d\mathbf{p}/dt$. Since $\mathbf{p}$ is the spatial part of a 4-vector whose fourth component is iE/c, $\mathbf{F}$ is clearly not part of a 4-vector. On the other hand, force density is $\mathbf{f} = \delta\mathbf{F}/\delta x_1\delta x_2\delta x_3 \rightarrow d\mathbf{p}/dx_1dx_2dx_3dt = icd\mathbf{p}/dx_1dx_2dx_3dx_4$ and since $dx_1dx_2dx_3dx_4$ is (as we have just seen) a scalar invariant, the force density transforms like $\mathbf{p}$ and is therefore the spatial part of a 4-vector whose fourth component is iw/c, where w is the rate of doing work per unit volume.

A3 *Density and Flux*

Suppose that q is an invariant scalar quantity, such as a number of particles or an electric charge, and we express it in terms of a density n so that, for example, the number of particles dq in a volume element dV is ndV or $dq=n\,dx_1dx_2dx_3$. Then, since dq must be a scalar, n cannot be a scalar. But $dq\,dx_4 = n(dx_1dx_2dx_3dx_4)$ and so n must transform like dx_4, in other words it is the fourth component J_4 of a 4-vector J_j.

Consider a system of particles, all at rest in a frame F' and characterised in this frame by a density n', which is the fourth component of a vector $J_j' = (\mathbf{J}', ic\,n')$ that has $\mathbf{J}'=0$ in this particular frame. In a frame F, in which F' has a velocity u along X_1, we will have $J_1 = -i\beta\gamma J_4' = \gamma u n'$, $J_2 = 0$, $J_3 = 0$, and $icn = \gamma(u) J_4' = \gamma(u) ic\, n'$, where $\beta = u/c$. Thus in F, $J_1=(u/ic)J_4 = u\,n$ or, in general, $\mathbf{J} = \mathbf{u}\,n$, and this is clearly the particle flux. Thus this flux and the particle density form a 4-vector

$$J_j = (\mathbf{J},\ ic\,n). \tag{A3.1}$$

The number of particles crossing an area element dx_2dx_3 in time dt is $dn = J_1dx_2dx_3dt = J_1dx_2dx_3dx_4/ic$, whose form shows that it is a scalar invariant. The scalar invariant equation $\partial J_k/\partial x_k = 0$, when written in three dimensional vector notation, is $\nabla.\mathrm{J} + \partial n/\partial t = 0$ which is a typical conservation, or continuity, equation for the quantity whose density is n.

The fact that a set of four components satisfies a law of this type does not guarantee that they are the components of a 4-vector. We cannot tell whether the zero on the r.h.s. of these equations is a scalar, a null vector or a null tensor. As an important example, in three dimensions the energy flux **N** and the energy density e must clearly satisfy $\nabla.\mathbf{N} + \partial e/\partial t = 0$, but the generalisation of this does not lead to a 4-vector $(\mathbf{N}, iec)$, instead, $(\mathbf{N}, iec)$ is row four of a tensor, the energy-momentum tensor. In terms of this tensor the conservation laws of energy and momentum can be combined in a single equation.

In three dimensions an equation such as $\nabla.\mathbf{J} + \partial u/\partial t = 0$ leads immediately to an integral equation

$$\iint J_n \mathrm{d}S + \iiint \partial u/\partial t \; \mathrm{d}V = 0 \qquad \text{(A3.2)}$$

which is why we recognise $\nabla.\mathbf{J} + \partial u/\partial t = 0$ as a conservation equation for the quantity with the density u. Before we can assert that this shows that $\iiint J_4 \mathrm{d}x_1 \mathrm{d}x_2 \mathrm{d}x_3$ is conserved, we need to know how to handle "3-surfaces" in a four dimensional "space".

A4 *Volume and 3-Surfaces*

In three dimensions the surface elements $\mathrm{d}r_1\mathrm{d}r_2$, $\mathrm{d}r_2\mathrm{d}r_3$, $\mathrm{d}r_3\mathrm{d}r_1$, although they are not invariant under rotations of the axes are at least quantities of the same kind. This is not true in four dimensions where we have mixed 3-surfaces e.g. $\mathrm{d}x_1\mathrm{d}x_2\mathrm{d}x_4 = ic\mathrm{d}x_1\mathrm{d}x_2\mathrm{d}t$ as well as purely spatial surfaces $\mathrm{d}x_1\mathrm{d}x_2\mathrm{d}x_3$. Moreover an interval $\mathrm{d}r_1^2 + \mathrm{d}r_2^2$ on an ordinary surface, or $\mathrm{d}x_1^2 + \mathrm{d}x_2^2 + \mathrm{d}x_3^2$ on a spatial "3-surface" is always positive, whereas on a mixed surface $\mathrm{d}x_1^2 + \mathrm{d}x_2^2 + \mathrm{d}x_4^2$ may be positive or negative, or even zero on the light cone.

Let x_j and y_j be two events on a particular 3-surface, the interval between them is $(x_j - y_j)^2$, which we can write as $r^2 - c^2t^2$ in terms of their spatial and time separations r and t. If, for all events on the surface $r > ct$, we call the surface spacelike, and there can be no causal connection between any two events on the surface. If, for all events on the surface, $ct > r$ the surface is timelike and all events on the surface could be connected by signals. This classification of a surface, or part of a surface is obviously relativistically invariant since it is made in terms of the invariant interval $r^2-c^2t^2$. The archetypal spacelike 3-surface is the 'NOW' volume, consisting of events all occurring NOW at the same time at different points in space.

Plane surfaces are important in ordinary geometry. A plane can be specified by a constant unit vector $\mathbf{n}$, its normal, and a point $\mathbf{r}_o$ on the plane. Any other point $\mathbf{r}$ on the plane satisfies $\mathbf{n}.(\mathbf{r}-\mathbf{r}_o) = 0$, and for any displacement $d\mathbf{r}$ on the plane $\mathbf{n}.d\mathbf{r} = 0$. An element of area on the plane can be written as $dS = \mathbf{n}.d\mathbf{S}$, and when the axes are rotated dS is an invariant scalar but the components of $d\mathbf{S}$ transform like those of a vector.

A 'plane' surface in space-time is similarly described in terms of a unit 'normal' n_j and an event $\underline{\underline{x}}_j$ in the 'plane' and any other event x_j on the surface satisfies the equation $n_j(x_j-\underline{\underline{x}}_j) = 0$ and again, for two nearby events on the surface, $n_j dx_j = 0$. A surface element ds_j of the plane can be expressed in terms of a scalar ds and the normal n_j as $n_j ds$, and both n_j and ds_j transform like vectors. Although this sounds very like the three dimensional case, a number of differences arise from the minus sign in the invariant element $n_j n_j$. If we write $n_j=(\mathbf{v}, iw)$ we have $n^2 = v^2 - w^2$, and a unit vector can have either $n^2 = +1$ or -1. If $n^2 = +1$ the normal is spacelike and lies outside the light-cone; if $n^2 = -1$ it is timelike and lies within the light-cone. If it is timelike, and w is positive, n_j points towards the future. Separations in the surface will satisfy $\mathbf{v}.d\mathbf{r} - wcdt = 0$ and so must have $dr > cdt$, making the surface spacelike. Suppose that $n_j=(0,0,0,i)$ which is clearly time-like, and the particular plane is $n_j x_j=0$, then $-ct = 0$ and the plane is just the ordinary 'now' space in three dimensions; on the other hand if $n_j x_j=-cT$ it is the ordinary three dimensional space at $t=T$.

If a surface is timelike, $dr < cdt$ on the surface, but we also have $\mathbf{v}.d\mathbf{r} - wcdt = 0$, which requires $v > w$, so that the normal must be spacelike. The normal to a timelike surface is spacelike, and a timelike normal defines a spacelike surface, but a spacelike normal does not necessarily define a timelike surface nor is the normal to a spacelike surface necessarily timelike.

In the laboratory frame the timelike normal $(0,0,0,ic)$ defines a plane 3-surface in which the element of 3-area is just the usual volume element $dx_1dx_2dx_3$. In the laboratory frame this has a definite value dV, but it is not an invariant scalar because it lacks the fourth component dx_4. We must now enquire how this description behaves under a Lorentz transformation.

With $x_j = (x_1, x_2, x_3, icT)$ the equation of a plane 3-surface is $n_kx_k + cT = 0$. With $n = (0,0,0,i)$ this describes the ordinary three dimensional space at $t = T$. If it is to be a plane spacelike surface in a frame F' moving along the x_1 axis with a velocity u it must be described in F' by $n_m'x_m' + cT' = 0$ with $n_m'n_m' = -1$ and T' a constant. This makes $n_1'\gamma(u)(x_1-uT) + in_4'\gamma(u)(cT-ux_1/c) + cT' = 0$ an identity in x_1, so that $n_1' = iun_4'/c$. Since $n_1'^2 + n_4'^2 = -1$ we get $n_1' = u\gamma/c$, $n_4' = i\gamma$ and $T' = (un_1'/c - in_4')\gamma T = T$. This is consistent with n_j being a 4-vector which transforms to n_j' in the usual way and with T being an invariant scalar, and so the spacelike 'now' plane in F becomes a spacelike plane in F'. In F the ordinary time T was also the proper time associated with the 'now' space in F and it remains the proper time associated the surface in F' whose normal is $n' = \gamma(-u/c, 0, 0, i)$. The normal has been 'tilted' in space-time.

The ordinary volume element $dx_1dx_2dx_3$ in the frame F has a definite value dV but is, as we have already remarked, not a scalar since it lacks the fourth component $dx_4 = ict$. We can however relate dV to a covariant vector $dV_j = v_jdV/ic$, where v_j is the 4-velocity $v_j = (0,0,0,ic)$ of a body at rest in the frame F (where $\gamma(v) = 1$ because $\mathbf{v} = 0$). This gives a covariant generalisation of the ordinary volume element and allows us to construct integrals of covariant quantities such as, for example, J_jdV_j, where J_j is the electric current density 4-vector.

A5 *Invariants of a Second-Rank Tensor*

Since, for a Lorentz transformation, $a^{-1}_{ij} = a_{ji}$ we see that for any second rank tensor $F_{ij}' = a_{im}a_{jn}F_{mn} = a_{im}F_{mn}a^{-1}_{nj}$. In matrix notation this is the similarity transformation aFa^{-1} which leaves the eigenvalues s_k of F unchanged. The eigenvalues s_k are the roots of the determinantal equation $|F_{ij} - s\delta_{ij}| = 0$ and, since they are unchanged the coefficients of the various powers of s, i.e. s^0, s^1, s^2, and s^3 in the determinant must also be unchanged (the coefficient of s^4 is automatically unity).

These scalar invariants are the traces of the matrix (tensor) F. A trace of order n is the sum of the principal minors of order n of F. A principal minor of order n is a determinant of that order which is formed from elements disposed symmetrically about the principal diagonal of F. The trace of order 1 is just the sum of the diagonal elements of F.

If we identify the principal minors of F by their diagonal elements, the four invariants are

$$I_1 = F_{11}+F_{22}+F_{33}+F_{44};$$
$$I_2 = F_{11},F_{22} + F_{22},F_{33} + F_{33},F_{44} + F_{11},F_{33} + F_{22},F_{44} + F_{11},F_{44};$$
$$I_3 = F_{11},F_{22},F_{33} + F_{22},F_{33},F_{44} + F_{33},F_{44},F_{11} + F_{44},F_{11},F_{22};$$
$$I_4 = F_{11},F_{22},F_{33},F_{44} = |F|. \qquad (A5.1)$$

A6 *The Electromagnetic Energy-Momentum Tensor*

The force density in charged matter is (in 3-dimensional notation) $f_\kappa = (\rho\mathbf{E}+\mathbf{j}\times\mathbf{B})_\kappa = \partial T^M_{\kappa\lambda}/\partial r_\lambda-(1/c^2)\partial N_\kappa/\partial t$, where N is Poynting's vector and T^M is Maxwell's stress tensor, defined in Eq.(8.11). The fourth component of the 4-vector force density is $f_4=iw/c=iE.j/c=-i(\nabla.N+\partial e/\partial t)$ where the energy density is $e=\frac{1}{2}(\varepsilon_0 E^2+B^2/\mu_0)$. It is fairly easy to see that these equations can be assembled as one 4-vector equation

$$f_i = - \partial T^{em}_{ij}/\partial x_j \ , \qquad (A6.1)$$

but much more tedious to show that the electromagnetic stress tensor T^{em}_{ij} can be expressed in terms of the field tensor F_{ij}, the scalar invariant, $E^2-B^2c^2$, and the invariant function δ_{ij} as

$$T^{em}_{ij} = -[F_{ik}F_{kj} - (E^2/c^2-B^2)\delta_{ij}]/\mu_0. \qquad (A6.2)$$

It is also possible to eliminate the invariant term containing the fields and express the stress tensor in terms of the field tensor alone as

$$T^{em}_{ij} = -(F_{ik}F_{kj}-\tfrac{1}{4}F_{kl}F_{lk}\delta_{ij})/\mu_0=(F_{ik}F_{jk}-\tfrac{1}{4}F_{kl}F_{kl}\delta_{ij})/\mu_0. \qquad (A6.3)$$

If the charged matter can be treated as a continuum and described by a kinetic energy-momentum tensor K_{ij} we will have

$$\partial K_{ij}/\partial x_j = f_i = -\partial T^{em}_{ij}/\partial x_j \qquad (A6.4)$$

and if we define a total energy momentum tensor

$$T_{ij} = K_{ij} + T^{em}_{ij} \tag{A6.5}$$

the equation

$$\partial T_{ij}/\partial x_j = 0 \tag{A6.6}$$

contains the complete laws of conservation of energy and linear momentum for a system of charged matter interacting through the e.m. field.

Because of this and because T_{ij} is symmetric the third rank tensor

$$Y_{ijk} = x_i T_{jk} - x_j T_{ik} \tag{A6.7}$$

satisfies

$$\partial Y_{ijk}/\partial x_k = 0 \tag{A6.8}$$

and so the quantity

$$J_{ij} = (1/ic)\iiint Y_{ij4}\, dx_1 dx_2 dx_3 = (1/ic)\iiint Y_{ij4}\, dV \tag{A6.9}$$

is a second rank tensor and a constant of the motion. Its spatial components correspond to the total angular momentum and its mixed space-time components describe the motion of the centre of mass.

The total angular momentum of the matter and the field are conserved and the velocity of the centre of mass of the whole system of matter and fields is constant.

Although the formulation of the energy, linear and angular momentum conservation laws in terms of the energy-momentum tensor is well adapted to discussing the dynamics of dilute plasmas in rather general terms, we might imagine that in dealing with, for instance, dielectric or conducting fluids it would be useful to formulate the theory in terms of the auxiliary vectors D and H and the constitutive relations between these vectors and E and B. However, this is not the case, fundamentally because there is no simple basic law that expresses matter forces within a polarisable medium in terms of the macroscopic field vectors alone. At the very least we need to know the dependence of the constitutive relations on strain or, alternatively, the rank-4 electrostrictive tensor, a difficulty that was well understood by Helmholtz over a century ago but then seems to have been forgotten until relatively recently. As Shockley (1968) and Robinson (1975) have shown, even in a cubic medium with a linear relation between D and E, which must be isotropic, the relation between the body-force and the fields is not isotropic, and therefore cannot be expressed in terms of E and D alone.

There are further difficulties in using the energy-momentum tensor even when dealing with structureless charged particles in vacuum. The energy flux $N=E\times B/\mu_0$, the momentum density N/c^2 and the energy density $\tfrac{1}{2}(\varepsilon_0 E^2+B^2/\mu_0)$ are a complete description of the dynamical properties of a radiation field well away from its sources, but in the vicinity of a charged body there are extra stresses in the field which cannot be ignored. There is an excellent account of these effects by Rohrlich (1965).

A7 Retarded Potentials

It is important that we prove that the retarded potentials satisfy the inhomogeneous wave equations for the vector and scalar potentials **A** and ϕ; for this is one of the results that shows most clearly that the velocity of propagation of electromagnetic effects in vacuum does not depend on the velocity of the source. It is also a result that does not depend in any way on an assumed waveform.

It suffices to show that the scalar retarded potential

$$\phi(\mathbf{R},T) = \iiint \rho(\mathbf{r},t)\,dV(\mathbf{r})/(4\pi\varepsilon_0|\mathbf{R}-\mathbf{r}|), \qquad \text{(A7.1)}$$

where

$$t = T - |\mathbf{R}-\mathbf{r}|/c, \qquad \text{(A7.2)}$$

satisfies the wave equation

$$\Box^2\phi = \nabla^2\phi - (1/c^2\partial^2\phi/\partial t^2 = -\rho/\varepsilon_0, \qquad \text{(A7.3)}$$

since the proof for **A** is almost identical.

The static solution of (A7.3) is

$$\phi(\mathbf{R}) = \iiint \rho(\mathbf{r})\,dV(\mathbf{r})/(4\pi\varepsilon_0|\mathbf{R}-\mathbf{r}|), \qquad \text{(A7.4)}$$

and we see that the singularity in (A7.1) at $\mathbf{r} = \mathbf{R}$, where the retardation can be ignored, contributes $-\rho(R,T)/\varepsilon_0$ to $\nabla^2\phi$. Away from this singular point we can change the order of integration and differentiation and so

$$\nabla^2\phi(\mathbf{R},T) = \iiint \nabla_R^2\{\rho(\mathbf{r},t-\xi/c)/4\pi\varepsilon_0\xi\}\,dV(\mathbf{r}) \qquad \text{(A7.5)}$$

where $\xi = |\mathbf{R} - \mathbf{r}|$. The function on which ∇_R^2 operates is, as far as its dependence on **R** goes, a scalar function of ξ the distance from an origin at **r** and so we can use the relation

$$\nabla^2 f(\xi) = \xi^{-2}\partial/\partial\xi(\xi^2\partial f/\partial\xi) = \xi^{-1}\partial^2/\partial\xi^2(\xi f).$$

This gives $\nabla^2\phi(\mathbf{R},T) = \iiint(1/4\pi\varepsilon_0\xi)\partial^2/\partial\xi^2\rho(\mathbf{r},T-\xi/c)\mathrm{d}V$ and so

$$\nabla^2\phi(\mathbf{R},T) = (1/c^2)\partial^2/\partial T^2\iiint\{\rho(\mathbf{r},T-\xi/c)/4\pi\varepsilon_0\xi\}\mathrm{d}V.$$

The charge in the immediate vicinity of the singularity at $\mathbf{r}=\mathbf{R}$ makes no contribution to ϕ or to $\partial^2\phi/\partial T^2$ and so this last result is $\nabla^2\phi = (1/c^2)\,\partial^2\phi/\partial T^2$. When we add the contribution from the singularity at $\mathbf{r} = \mathbf{R}$ we get Eq.(A7.3).

There is also an 'advanced' potential solution of the wave equation, and the reader interested in the physical significance of this solution and its relation to causality, should consult Dirac (1938).

A8 *Linearity of the Lorentz Transformation*

Some years ago, when I first thought of writing this book, I was very surprised that I could not find a text or monograph, however fat, that seemed to think that it was necessary to prove that the Lorentz transformation must be linear. Indeed few of them even bothered to mention that they were assuming that it was linear. I therefore tried to construct a proof for myself, and my initial effort was both clumsy and long-winded. A few years later I acquired a copy of 'The Theory of Space, Time and Gravitation' by Fock (1959), and at last found another author who had experienced the same difficulty and had also constructed a proof very similar to mine in both its content and its length.

I then remembered that even longer ago, when I was designing electron optical systems, I had read a discussion of the most general form of the relation between images and objects in optical systems and this, it seemed to me, was a very similar problem. After many false starts I finally tracked it down to Drude's 'The Theory of Optics' (1900). Drude discusses the algebraic transformation that maps objects in the 'object space' of an optical instrument to images in its 'image space', in more or less the following terms. If, to every point x_m in the object space, there corresponds one and only one point x_m' in the image space and vice versa, then it must be possible to use the formula to calculate x_m' from x_m and x_m from x_m' without any ambiguity, such as might arise if one of the relations were quadratic.

The most general algebraic relation with this property is

$$x_m' = \Sigma_n (a_{mn}x_n+b_m)/(c_n x_n+d), \qquad (A8.1)$$

where m and n run over 1,2,3. Clearly the same argument holds if m and n run over 1,2,3,4 with $x_4 = t$ or ict. But Eq.(A8.1) maps a plane, for which $c_n x_n + d = 0$, at infinity. This is acceptable in optics, as we expect the focal plane to be mapped at infinity, but it is not acceptable in relativity to have nearby points mapped at either an infinite distance, or at an infinitely remote time. In fact, as Drude points out, it is also not acceptable in telescopic systems. Thus the relation between x' and x can only be linear, with $x_m' = \Sigma_n (a_{mn}x_n+b_m)$ and, if we ignore the trivial effects of a displacement of the origin, we can reduce this to $x_m' = \Sigma_n a_{mn} x_n$.

A9 *Free particles and Newton's First Law of Motion*

At the time that the theory of special relativity was being formulated, theoretical physicists, especially in Austria and Germany (see the bibliography), were much concerned with questions of epistemology. Did physicists have an intuitive picture of what they meant by an inertial frame of reference, or should they try to give an operational definition of all their terms and concepts?

With respect to what, is the motion of a free particle rectilinear and of constant velocity? We can hardly answer 'with respect to an inertial frame' if we define an inertial frame as a frame relative to which free particles obey Newton's first law. We could perhaps do a little better by stating this law in terms of the relative motion of two free particles, but then we have to say what we mean by free particles. Do we appeal to our experience that the interaction between particles diminishes rapidly with their separation, and thus claim that a free particle is a particle so far from all others that its interactions can be ignored. Since Ernst Mach supposed that inertia itself was due to the effects of the distant stars, we might soon lose ourselves in a fog of cosmology.

I have postponed raising these questions until the very end of the book, since it seemed to me that most readers would already have an intuitive understanding of terms such as 'coordinate system', of concepts such as 'space' and 'time' associated with Newton's first law, and of the pictures that physicists use to describe the physical world.

Though there is something to be said for scrutinising our basic assumptions, the final test of our models is how far they enable us to interpret and predict experience, and ultimately of course how far they are found to be useful by engineers. We do not believe electromagnetic theory to be correct because we find the inverse square law of electrostatics intellectually and logically compelling; but because we have not so far come across phenomena with which it disagrees and because, in everyday life, we see innumerable examples of effects that can be interpreted, and in some cases (such as radio) were predicted by this theory. This need not stop us from introducing students to electromagnetic theory via a discussion of electrostatics, Ampère's Law and Faraday's Law of electromagnetic induction.

A10 *Some Physical Constants*

Velocity of light,	$c = 2.9979\times10^{8}\ \mathrm{ms^{-1}}$.
	$\mu_o = 4\pi\times10^{-7}\ \mathrm{H\ m^{-1}}$,
	$\varepsilon_o = 8.854\times10^{-12}\ \mathrm{Fm^{-1}}$.
Elementary charge,	$e = 1.6022\times10^{-19}$ C.
Electron mass,	$m = 9.1095\times10^{-31}$ kg.
Proton mass,	$M = 1.6726\times10^{-27}$ kg.
Planck's constant,	$h = 6.626\ \times10^{-34}$ J/Hz, $\hbar \equiv h/2\pi = 1.0546\ \times10^{-34}$ J/Hz.
Bohr radius,	$a_o = 5.292\times10^{-11}$m.
Bohr magneton,	$\beta\ = 9.274\times10^{-24}$ J/T.
Avogadro's number,	$N = 6.022\ \times10^{23}$ /gram mol.
Boltzmann's constant,	$k = 1.381\ \times10^{-23}\ \mathrm{J\ K^{-1}}$.
Stefan's constant,	$\sigma = 5.67\times10^{-8}\ \mathrm{Wm^{-2}K^{-4}}$.

Astronomical unit (mean radius of earth's orbit) 1.5×10^{8} km.

Gravitational constant, $G = 6.672\times10^{-11}\ \mathrm{Nm^2kg^{-2}}$.

Earth's radius, 6371 km.

Sun's radius, 695950 km.

REFERENCES and BIBLIOGRAPHY

Abraham M. (1905) '*Theorie der Electrizität*' **II** (Springer, Leipzig).

Aitchison I.J.R. & Hey A.J.G. (1982) '*Gauge Theories in Particle Physics*' p21 (Adam Hilger).

Alvager T. Bailey J.M. Farley F.J.M. Kjellman J. & Wallin I. (1964/5) *Physics Lett.* **12,** 260, *Arkiv.f.Fys.* **31**, 145.

Bailey J.M. Borer K. Combley F. Drumm H. Krièman F. Lange F. Picasso E. von Ruden W. Farley F.J.M. Field J.H. Flegel W. & Hattersley P.M. (1977) *Nature* **268,** 301.

Bradley J. (1728) *Phil.Trans.***35,** 637.

Brecher K. (1977) *Phys.Rev.Lett.* **39,** 1051.

Brillouin L. (1960) '*Wave Propagation and Group Velocity*' (Academic Press).

Dicke R.H.(1965) '*The Theoretical Significance of Experimental Relativity*' (Gordon and Breach).

Dirac P.A.M.(1938) Proc.Roy.Soc. **A167,** 148.
1958 '*The Principles of Quantum Mechanics*' (Oxford).

Drude P. (1900) '*The Theory of Optics*' (Dover Reprint).

Einstein A. (1905a) *Ann.der Physik* **17** 891, (1905b) *Ann.der Physik* **18,**639. See also Einstein, Lorentz, Minkowski and Weyl, '*The Principle of Relativity*',(Dover Reprint).

Essen L. & Parry J.V.L. (1958) *Phil.Trans.* **250,** 45.

Fock V. (1959) '*The Theory of Space, Time and Gravitation*' (Pergamon)

Gordon J.P. Zeiger W. and Townes C. H. (1955) *Phys.Rev.* **105,** 762.

Hafele J.C. & Keating R.E. (1972) *Science* **177,** 166. See also Schlegel R. (1974) *Am.J.Phys.* **42,** 183.

Heitler W. (1935) and subsequent editions, '*The Quantum Theory of Radiation*' Ch.1, (Oxford).

Isaac G.R.(1970) *Phys.Bull*.pp255-7

Ives H.E. & Stilwell G.R.(1938) *J.Opt.Soc.Am.* **28,** 215 and **31,** 369.

Jackson J.D.(1975) '*Classical Electrodynamics*' (Wiley).

Jaseja T.S. Javan A. Murray J. and Townes C.H. (1964) *Phys. Rev.* **133** A1221-1225

Kaufmann W.(1902) *Phys.Zeits.* **4,** 54.

Landau L.D. and Lifshitz E.M. (1960) '*Mechanics*' (Pergamon).

Lewis G.N.(1908) *Phil.Mag.* **16**, 705.

Liénard A.(1898) *L'Éclairage électrique* **16**,5,53,106.

Lindsay R.B. and Margenau H.(1936) '*Foundations of Physics*' (Wiley).

Lorentz H.A.(1903) *Proc. Amst. Acad.* (English edn.) **6**, 809. See also Lorentz (1909) '*Theory of Electrons*', (Dover reprint 1952), and under Einstein (1905) above.

Luchini P. and Motz H.(1990) ' *Undulators and Free Electron Lasers*' (Oxford).

Madey J.M.J.(1971) *J.Appl.Phys.* **42**, 1906.

Michelson A.A. & Morley E.W.(1887) *Am.J.Sci.* **34**, 333.

Møller C.(1971) '*The Theory of Relativity*', (Oxford).

Motz H.(1951) *J. Appl. Phys.* **22**, 527.

Motz H.(1982) pp 355-361 in '*Ludwig Boltzmann Gesammtausgabe*', see also Wagner S. pp 343-354. (Friedr. Vieweg & Sohn, Braunschweig/Wiesbaden).

Muirhead H. (1973) '*The Special Theory of Relativity*' (Macmillan)

Noether E.(1918) *Nachr.Akad.Wiss. Göttingen* **II** *Math.Phys.***K1** 235.

Nye J.F.(1957) '*Physical Properties of Crystals*' Ch.5 (Oxford).

Poincaré H.(1901) and (1904) see Whittaker **II**,p30, in the bibliography.

Podolsky B.and Kunz K.S. (1969) '*Fundamentals of Electrodynamics*' (Dekker).

Pound R.V.and Rebka G.A.(1960) *Phys.Rev.Lett.***4**, 274.

Pound R.V.and Snider J.L.(1964) *Phys.Rev.Lett.* **13**, 539.

Robinson F.N.H. (1971) Physica **54**, 329, (1973) '*Macroscopic Electromagnetism*' (Pergamon), (1975) *Physics Reports* **Vol. 16C**, no.6, 313-354.

Rohrlich F.(1965) '*Classical Charged Particles*' (Addison Wesley).

Rosen G.(1969) '*Formulations of Classical and Quantum Dynamical Theory*' (Academic Press).

Rosser W.G.V.(1968) '*Classical Electromagnetism via Relativity*' (Butterworth).

Shockley W.(1968) *Proc.Nat.Acad.Sci.***60**, 807.

Thomas L.W. (1927) *Phil.Mag.*(7) **3**,1.

Trester J.J. (1989) *Am.J.Phys.* **57**, 86.

Wapstra A.H.(1972) '*Atomic Masses and Fundamental Constants*' p.283 (Plenum).
Wiechert E. (1900) *Archives Néerlandaises* **5**,549.
Yourgrau W. and Mandelstam S. (1968) '*Variational Principles in Dynamics and Quantum Theory*' (Dover Reprint).

BIBLIOGRAPHY

Of the many books on relativity here are a few that I have found useful. Some of them have already been mentioned in the text.

Bowler M.G. '*Gravitation and Relativity*' (Pergamon 1976).

This concise account of general relativity is especially valuable for the vigour with which it demolishes the fallacy that general relativity is needed to deal with acceleration.

Dicke R.H. '*The Theoretical Significance of Experimental Relativity*' (Gordon and Breach 1965).

With its emphasis on experimental facts, this splendid book blows like a gust of fresh air through clouds of dogma and speculation.

Fock V. '*The Theory of Space, Time and Gravitation*' translated by N.Kemmer,(Pergamon 1959)

It contains several excellent chapters on special relativity.

Hagedorn R. '*Relativistic Kinematics*' (Benjamin 1964).

This is best summarised by the author's own subtitle 'A guide to the kinematic problems of high-energy physics.'

Møller C. '*The Theory of Relativity*' (Oxford 1971)

Heavy going but sound and sensible.

Muirhead H. '*The Special Theory of Relativity*' (Macmillan 1973).

In this excellent book the author "has tried to set out the exploitation of the principle of Lorentz invariance in the way in which it is done in high-energy physics".

Whittaker E.T. '*History of the Theories of Aether and Electricity*' (Nelson 1953).

Volume I ends in 1900 just before Planck, Lorentz, Poincaré, Einstein, Thomson and Rutherford inaugurated the era of what we still regard as 'MODERN' physics. Volume II deals with the first quarter of the twentieth century, the age of relativity, quantum mechanics, the electron and the nuclear atom. This is a unique, scholarly and comprehensive history of physics and, because Whittaker was a first-rate applied mathematician, the physics from the Greeks to Dirac is treated professionally. If he gave Einstein less than his due in connection with special relativity, it was probably a reaction to the neglect of Poincaré and Lorentz, at the time that Whittaker was writing. He was more generous about general relativity, photoelectricity and Brownian motion. For physicists, the two volumes must be one of the best detective stories ever written, relentlessly pursuing the central theme of physics over two millenia.

For the reader interested in the intellectual climate from which special relativity emerged, an important text is '*Ludwig Boltzmann Gesammtausgabe*', edited by Roman Sexl and John Blackmore,(Friedr.Vieweg&Sohn, Braunschweig/Wiesbaden). There is no question that Boltzmann had a profound influence on both Einstein and Planck.

Finally, although most physicists accept the basic validity of special relativity, or at least the Lorentz transformation, some physicists find all the talk about clocks and the twin paradox hard to ignore. A useful guide to this penumbral world is '*THE SPECIAL THEORY OF RELATIVITY, a critical analysis*' by L Essen (O U P 1971).

Essen is not primarily concerned to deny the validity of special relativity, but rather to expose the weakness of many of the arguments advanced in its support. As the inventor of the caesium clock he has rather better credentials than most critics of relativity.

ANSWERS TO PROBLEMS

1.1 The time delay from A to B at the velocity of light is 1msec and thus a phase delay of 18° at 50 Hz, if A follows B then B will receive a signal lagging by 36° and so on.

2.1

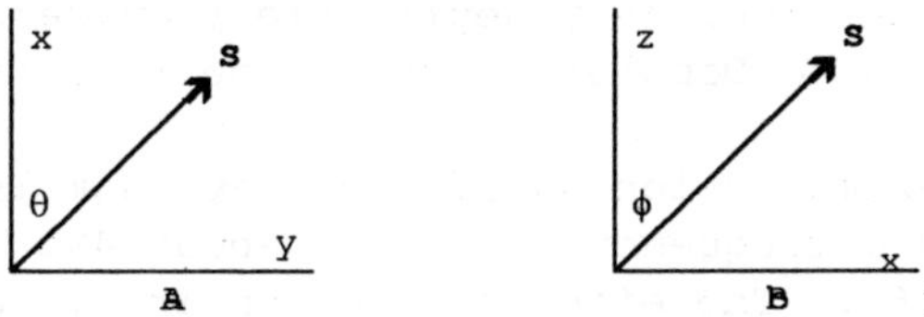

1

In Figure A $d\theta = \omega_z dt$, $S_x = S'\cos\theta$ and $S_y = S'\sin\theta$ so $dS_x = -S'_y \omega_z dt$, similarly in Figure b $S_x = S'\sin\phi$ and $dS_x = S'_z \omega_y dt$, and so $dS_x = (S'_z \omega_y - S'_y \omega_z) dt = (\boldsymbol{\omega} \times \mathbf{S}')_x dt$ etc. $d^2S/dt^2 = \boldsymbol{\omega} \times \mathbf{v}$.

2.2 The second event occurs at $x' = \gamma(u) x_2 = 10\sqrt{(9/8)} = 10.61$km and at $t' = -\gamma u x_2/c^2 = -11.8$ μsec i.e before the event at the origin.

2.3 It lasts 10^9 secs, since $\gamma = 1/\sqrt{(1-10^{-10})}$ the rocket clock will read 1/20 of a second less than 10^9 secs.

2.4 The 4 values of γ are 2.3, 7.1, 22.4, 70.7 and it travels 6.2, 21.1, 67.2, 212 $\times 10^{-10}$m.

2.5 $36/13 \times 10^8$ m/sec. 2.6 0.9998c. 2.7 (0, 0.58c, 0).

2.8. Change the axes so that x is parallel to **u** and **v** then $u = c/\sqrt{2}$ $v = c/\sqrt{2}$ and the particle remains at the origin of F'.

2.9 Change to a frame F' with a velocity u in which the particle is stationary and after it breaks up the two fragments have velocities w and -w, their relative velocity is $2w/(1+w^2/c^2) = v = 2c/3$ and $w/c = 3/2 - \sqrt{(5/4)}$, the velocities relative to C are $(u+w)/(1+uw/c^2)$ and $(u-w)/(1-uw/c^2)$. 0.84c and 0.38c

2.10 $d\gamma = (1-u^2/c^2)^{-3/2}\, udu/c^2 = \gamma^3 udu/c^2$ and $udu = \frac{1}{2}d(u^2) = \frac{1}{2}d(\mathbf{u}.\mathbf{u}) = \mathbf{u}.d\mathbf{u}$.

3.1 The path traversed by the reflected wave is reducing at 4000km/hr =1111m/sec = 3.7×10^{-6}c so the frequency shift is 50×3.7=185kHz.

3.2 The shell is nearest to the plane when it has no velocity component towards the plane and so, apart from the completely negligible transverse effect, there is no Doppler shift of the reflected signal.

3.3 The maximum velocity of the diaphragm is 4π m/sec giving a frequency wobble of about 4%. A semitone is about 6%. The effect is not entirely negligible.

3.4 In the laboratory frame the frequency is u/ℓ, in the ion frame it is γ times greater. $5.10^{11}=10^3 u\gamma(u)$, ∴ u=0.86c. $\gamma \approx 2$ and, if the ion were say Mg, the required energy would be about 50GeV.

3.5 0.6c.

3.6 Frequency ratio $[(1-1/3)/(1+1/3)]^{\frac{1}{2}}/[(1-1/2)/(1+1/2)]^{\frac{1}{2}} = [3/2]^{\frac{1}{2}}$, intensity ratio 3/2.

3.7 Velocity 0.9c, γ=2.3, elapsed proper time is 0.44ns.

3.8 Bohr radius $r=5.29.10^{-11}$m, angular momentum $m\omega r^2=\hbar$, $m=9.11\times10^{-31}$, $\hbar=1.055\times10^{-34}$ ∴ $\omega=4.14\times10^{16}$. $\Omega_T = \mathbf{a}\times\mathbf{v}/2c^2 = \omega^3 r^2/2c^2 = 1.10\times10^{12}\text{sec}^{-1}$.

3.9 0.095c.

3.10 γ =5/3, $x'=\gamma$ (0-ut)=-5/3(2.4/10)=-0.4m.

3.11 E ~ 2.5mV/m, $d\nu/\nu \sim 6\times10^{-26}$.

3.12 From the stellar aberration we get the earth's orbital velocity $v=10^{-4}$c, combining this with the length of the year we get the radius of the earth's orbit

$R=1.5\times10^{11}$m. and from the angular diameter of the sun we get the radius of the sun as $r = 4.65\times10^{-3}R$. The acceleration of the earth towards the sun, due to gravity, is $g'=v^2/R = 6\times10^{-3}$m/sec². Finally we get $\delta\nu/\nu = -g'/c^2 \int_r^R (R/x)^2 dx = -g'R/c^2[R/r -1] = -2.14\times10^{-6}$.

3.13 Since, at temperatures below about 10,000°, atomic velocities are negligible compared with c the observed wavelength from an atom receding with a velocity v is well enough approximated by $\lambda=\lambda_o(1+v/c)$ and $d\lambda=\lambda_o dv/c$. The number of atoms with velocities between v and $v+dv$ is proportional to $\exp[-Mv^2/2kT]dv$ and thus with $v= c(\lambda-\lambda_O)/\lambda_O$ we get the quoted result. The mean square line width is $\delta\lambda^2=\lambda^2 kT/Mc^2$ giving $\delta\lambda/\lambda \sim 1.3\times10^{-6}$ to be compared with $0.6/589 \sim 10^{-3}$.

3.14 $\nu \sim 3.6\times10^{18}$, the height was 22m and $gH/c^2 = 2.4\times10^{-15}$ and so $\delta\nu \sim 8.6$ kHz.

3.15 Suppose the length of the beam is L then in the rest frame of the electrons the elapsed time is $L/[v\gamma(v)]$ and the time that the repulsive force can act is reduced very greatly, thus for 1GeV electrons the reduction from L/c is 1/2000. The reader may be able to suggest a number of other methods of obtaining the same result.

4.1 In terms of their components we have dE/dp_λ, and since E does not depend on direction we can choose one of the axes as in the direction of p and then the only component of v is along this axis.

4.2 $m d\dot{r}/dt=-e(E+r\dot{\theta}B)$ and $d(mr^2\dot{\theta})/dt = er\dot{r}B$, so that $d/dt[\frac{1}{2}r^2eB-mr^2\dot{\theta}]=0$. At the cathode $\dot{\theta}=0$ and $L^*=\frac{1}{2}a^2eB$, thus at the anode $mb^2\dot{\theta}=\frac{1}{2}(b^2-a^2)eB$, $\dot{\theta}=\frac{1}{2}eB/m(b^2-a^2)/b^2$ and the electron k.e. is $(e^2B^2/8m)(b^2-a^2)^2/b^2$. This must be less than eV the decreased potential energy. The cavity magnetron invented by Randall, Sayers and Boot in 1942 is still, because of its efficiency, the most important generator of microwave power.

4.3 $4(2mE)^{1/2}/k$. 4.4 $g/3$.

5.1 $r^2[\tfrac{1}{2}eB-\gamma(r\dot\theta)m_o\dot\theta^2]$.

5.2 $\gamma = 1/(1-0.64)^{1/2} = 1/0.6 = 5/3$.

5.3 $EdE=c^2pdp$, $v=dE/dp=cp/\sqrt{(m^2c^2+p^2)}$, $v^2/c^2=p^2/(p^2+m^2c^2)$, $1-v^2/c^2 = m^2c^2/(p^2+m^2c^2)$, $\gamma = \sqrt{(1+p^2/m^2c^2)}$, $dv/dp = m^2/(m^2+p^2/c^2)^{3/2} = 1/m\gamma^3$. $\therefore$ $m^* = m\gamma^3$

5.4 $\gamma = 3$, $v = c\sqrt{(8/9)}$.

5.5 $pc = \sqrt{(E^2-m^2c^4)} = E\sqrt{(1-1/4\times10^6)}$, $E/pc \sim 1+1/8\times10^6$.

5.6 Electrons which cross at zero voltage have $\gamma = 2$, $v^2/c^2 = 3/4$ and $v=2.598\times10^8$ m/sec, those which cross at max. voltage have $\gamma = 2.02$ and $v' = 2.607\times10^8$ m/sec, $L/v - L/v' = 10^{-10}/4$, $L \sim 2$m. A neater method uses $cd(1/v) = -(mc^2/E)^2[1-(mc^2/E^2)^2]^{-3/2}\,dE/E$.

5.7 approx. 1.4×10^{-15}m.

5.8 $\gamma=1.25$, kinetic energy$=0.25Mc^2=2.25\times10^{16}J=2.25\times10^7$GWsec = 260GWdays. At 5p/kWhr the cost is £315.5 million.

5.9 $\omega = eB/\gamma m$, $1-\omega^2r^2/c^2 = 1/\gamma^2$, $r\sim0.17$m.

6.1 About 4μm. Note that the block's length is irrelevant. Thus micron size objects can be moved by their own length. This is used in optical tweezers. (A Ashkin and J M Dziedzic *Phys. Rev. Lett.* **30**, 139, 1973)

6.2 Gas $\sim 3.10^4$ N/m², radiation $\sim 3.10^{-4}$ N/m².

6.3 When $\theta=90°$ Eq. (6.16b) can be written in terms of the photon energies as $1/E_2-1/E_1=1/mc^2$ giving $E_2=100/1.2 = 83.3$keV so that the electron K.E. must be 16.7keV.

6.4 In the zero momentum frame the momentum carried off by radiation must be zero, but it must also carry off energy. For a single photon $E=pc$.

6.5 Direct substitution in Eq.(6.15) gives $E = 1.5Mc^2$.

6.6 A first integration gives $v^2=(Ft/m)^2(1-v^2/c^2)$ so that $dx=vut=Fut^2/2m[1+(Ft/mc)^2]^{\frac{1}{2}}$ and this then integrates to give $x=(mc^2/F)[(1+F^2t^2/m^2c^2)^{\frac{1}{2}}-1]$. See discussion following Eq.(6.30).

7.1 To first order in infinitesimals

$$\begin{matrix} 1 & -d\phi & 0 \\ d\phi & 1 & -d\theta \\ 0 & d\theta & 1 \end{matrix} .$$

7.2 $F_aG_aa^2+F_bG_bb^2+F_cG_cc^2+(F_aG_b+F_bG_a)\mathbf{a.b}$ + etc.

7.3 Both u_α and $\partial/\partial x_\beta$ are vectors and so $\partial u_\alpha/\partial x_\beta$ is a 2nd rank tensor and clearly $\sigma_{\alpha\beta}$ is symmetric.

7.4 Because $\boldsymbol{\sigma}_{\alpha\beta}$ is symmetric it can be referred to its principal axes taking the form diag$(\boldsymbol{\sigma}_{11},\boldsymbol{\sigma}_{22},\boldsymbol{\sigma}_{33})$. The sides of a unit cube become $1+\boldsymbol{\sigma}_{11}$, $1+\boldsymbol{\sigma}_{22}$, $1+\boldsymbol{\sigma}_{33}$, and, since the $\boldsymbol{\sigma}$'s are small the fractional change in volume or dilatation is $\boldsymbol{\sigma}_{11}+\boldsymbol{\sigma}_{22}+\boldsymbol{\sigma}_{33}$, which is the invariant trace of the strain tensor. [see e.g. Landau and Lifshitz 'Theory of Elasticity' pp1-3, (Pergamon 1970)].

7.5 See appendix: each of the terms in I_1 and I_3 is individually zero.

7.6 In the order the terms are listed in Eq.(A5.1) we get $M^2{}_3 + M^2{}_1 + N^2{}_3 + M^2{}_2 + N^2{}_2 + N^2{}_1$.

8.1 Assume that the waves propagate along the z axis. As they are plane waves the x and y derivatives of any field component are zero. It follows from the two divergence equations that their z components are zero. The two curl equations give two pairs of equations for the two independent transversely polarised waves $\dot{B}_x=\partial E_y/\partial z$, $\mu_0\varepsilon_0\dot{E}_y=\partial B_x/\partial z$ and $-\dot{B}_y=\partial E_x/\partial z$, $\mu_0\varepsilon_0\dot{E}_x=-\partial B_y/\partial z$.

8.2 $\omega^2=\beta^2c^2+(\pi c/2a)^2$. $B_x=-(\beta/\omega)E_y$, $B_z=-i(\pi/2\omega a)E_o\sin\pi x/2a$. Clearly $\omega/\beta>c$ and, since $\omega/\beta\, d\omega/d\beta = d\omega^2/d\beta^2 = c^2$, $d\omega/d\beta<c$.

8.3 See Appendix.

8.4 In cylindrical coordinates $E_r= V/r\ln b/a$ where a and b are the radii of the inner and outer conductors. The magnetic field $\mathbf{H}=\mathbf{B}/\mu_o$ has the sole component $H_\theta=I/2\pi r$ so that the only component of Poynting's vector is $N_z= IV/2\pi r^2\ln b/a$ and the power flow $\int N\,2\pi r dr = IV$.

8.5 HINT: you will have to obtain the form of the stress tensor at the surface of the enamel and calculate the resulting force acting over the entire surface of the enamel and add this to IB.

9.1 Consider a wave with only the transverse components E_2 and $B_3=E_2/c$; then Eqs.(9.14a,b) yield
$B_3'=\gamma(B_3-uE_2/c^2) = \gamma(1-u/c)B_3$ and
$E_2'= \gamma(E_2-uB_3)= \gamma(1-u/c)E_2 = [(1-u/c)/(1+u/c)]^{1/2}E_2$,
with a similar result for B_3'. As the wave power is E_2B_3/μ_o it is reduced by a factor $(1-u/c)/(1+u/c)$.

9.2 In Eqns.(14a,b) replace **E** by **P** and **B** by **M**.

9.3 The time average value of the only component of the momentum density is $\langle p_3\rangle= \varepsilon_o E_o^2/c$, and the integrals over dr_1dr_2 of the components $l_1=x_2p_3$ and $l_2=-x_1p_3$ vanish by symmetry.

9.4 The time average of the momentum component p_1 is $\langle p_1 \rangle = \frac{1}{2}\Re e(E_2B_3^*-E_3B_2^*)/\mu_o c^2$ and after some algebra this gives $\langle p_1 \rangle = (\varepsilon_o/\omega)E_0\partial E_0/\partial x_2$ and similarly, $\langle p_2 \rangle = -(\varepsilon_o/\omega)E_0\partial E_0/\partial x_1$. The component $\langle l_3 \rangle = x_1\langle p_2) - x_2\langle p_1 \rangle$ of the angular momentum density is therefore

$$\langle l_3 \rangle = -(\varepsilon_o/\omega)[x_1E_0\partial E_0/\partial x_1 + x_2E_0\partial E_0/\partial x_2].$$

and the angular momentum per unit length is

$$L_3 = -(\varepsilon_o/\omega)\iint[x_1E_0\partial E_0/\partial x_1 + x_2E_0\partial E_0/\partial x_2]dx_1dx_2.$$

The energy per unit length is $U = \varepsilon_o\iint E_0^2 dx_1dx_2$ and this can be expressed as the sum of a term

$$\tfrac{1}{2}\varepsilon_o\iint E_0^2 dx_1dx_2 = \tfrac{1}{2}\varepsilon_o\{[E_0^2x_1]_0^\infty - 2\iint[x_1E_0\partial E_0/\partial x_1]dx_1dx_2\}$$

and a similar term with x_1 replaced by x_2. The integrated parts are both zero, for E_0 vanishes at ∞. Thus finally we obtain $U = \omega L_3$, and a familiar result in quantum mechanics.

9.5 1.79×10^{-8}. About 10^7 (vacuum) wavelengths or 5m.

9.6 2.7×10^{14}V/m, enough to strip all the electrons off most atoms.

9.7 About 0.2mm/sec. 0.02T.

9.8 Let the electron velocity be v parallel to the 1 axis. In the electron rest frame F there will be electric field components E_2 and E_3 but no magnetic field. In the laboratory frame with a velocity $u=-v$, Eqs. (14a,b) give $E_2'=\gamma E_2$, $E_3'=\gamma E_3$, $B_2'=-\gamma vE_3/c^2$, $B_3'=\gamma vE_2/c^2$. The components of the force are $F_2' = e(E_2'-vB_3') = eE_2/\gamma$ and $F_3' = eE_3/\gamma$, which both tend to zero as $v\to c$.

10.1 The force per unit volume is $f_\mu = -\partial\pi_\mu/\partial r_\mu$ and $q_\lambda=\rho v_\lambda$. The momentum equation gives

$$v_\lambda\partial\rho v_\mu/\partial r_\mu + \rho v_\mu\partial v_\lambda/\partial r_\mu + v_\lambda\partial\rho/\partial t + \rho\partial v_\lambda/\partial t = f_\mu.$$

Since mass is conserved the 1st and 3rd terms combine to give zero, while the 2nd and 4th terms combine to give the total, or hydrodynamic, derivative

$$d/dt=v_\mu\partial/\partial r_\mu+\partial/\partial t.$$

10.2 See Eq.(10.4). The pressure π is the momentum flux across a fixed plane and, for this isotropic system, it is $K_{11}=\nu<E-mc^2/\gamma>/3 = \nu<E(1-1/\gamma^2)>/3$. For $v<<c$ this becomes $\pi= n<Ev^2/c^2>/3$ and with $<E>=mc^2$ we get the classical result $\pi=\nu m<v^2>/3$ or $2<T>/3$ where T is the kinetic energy per particle. The extreme relativistic case (applicable to radiation) gives $\pi=<\mathcal{E}>/3$ where $\mathcal{E}$ is the energy density.

11.1 (i) $L\ddot{q} + q/C = 0$. (ii) $p=L\dot{q}$. (iii) $H=p^2/2L +q^2/2C$, $\dot{p}=-q/C$, $\dot{q}=p/L$.

11.2 $p=m_0 v/(1-v^2/c^2)^{1/2}$, $v^2=p^2c^2/(m_o{}^2c^2+p^2)$, $H=pv-L=m_0c^2(1+p^2/m_0{}^2c^2)^{1/2}$.

11.3 Eq.(11.31) is $dp_i/dt=ev_\mu\partial A_\mu/\partial q_i-e\partial\phi/\partial q_i$, and the fourth component is $dE/dt=-e\mathbf{v}.\partial\mathbf{A}/\partial t+e\partial\phi/\partial t=e\mathbf{v}.\mathbf{E}+e[\mathbf{v}.\nabla\phi+\partial\phi/\partial t]=e[\mathbf{v}.\mathbf{E}+d\phi/dt]$.